Essential Clinically Applied Anatomy of the Peripheral Nervous System in the Head and Neck

Essential Clinically Applied Anatomy of the Peripheral Nervous System in the Head and Neck

Paul Rea
MBChB, MSc, PhD, MIMI, RMIP, FHEA, FRSA
University of Glasgow, Glasgow, UK

ELSEVIER

AMSTERDAM • BOSTON • HEIDELBERG • LONDON
NEW YORK • OXFORD • PARIS • SAN DIEGO
SAN FRANCISCO • SINGAPORE • SYDNEY • TOKYO

Academic Press is an imprint of Elsevier

Academic Press is an imprint of Elsevier
125 London Wall, London EC2Y 5AS, UK
525 B Street, Suite 1800, San Diego, CA 92101-4495, USA
50 Hampshire Street, 5th Floor, Cambridge, MA 02139, USA
The Boulevard, Langford Lane, Kidlington, Oxford OX5 1GB, UK

Notices
Knowledge and best practice in this field are constantly changing. As new research and
experience broaden our understanding, changes in research methods, professional practices,
or medical treatment may become necessary.

Practitioners and researchers must always rely on their own experience and knowledge in
evaluating and using any information, methods, compounds, or experiments described herein.
In using such information or methods they should be mindful of their own safety and the safety
of others, including parties for whom they have a professional responsibility.

To the fullest extent of the law, neither the Publisher nor the authors, contributors, or editors,
assume any liability for any injury and/or damage to persons or property as a matter of products
liability, negligence or otherwise, or from any use or operation of any methods, products,
instructions, or ideas contained in the material herein.

ISBN: 978-0-12-803633-4

Library of Congress Cataloging-in-Publication Data
A catalog record for this book is available from the Library of Congress

British Library Cataloguing-in-Publication Data
A catalogue record for this book is available from the British Library

For information on all Academic Press publications
visit our website at http://store.elsevier.com/

ELSEVIER Book Aid International

Working together
to grow libraries in
developing countries

www.elsevier.com • www.bookaid.org

CONTENTS

LIST OF FIGURES

LIST OF TABLES

PREFACE

Anatomy textbooks tend to be crammed with highly complex detail with clinical applications placed in boxes as an after-thought, and not part of the main text. The opposite is true for clinical resources where the anatomy tends not to be examined in much detail, and it is difficult to relate the anatomy to the clinical relevance and associated patholo-gies. In addition, these resources tend to be bulky texts and with infor-mation scattered throughout making access difficult to essential key facts and clinical applications.

Therefore, the purpose of this resource in Clinically Applied Anatomy of the Nerves in the Head and Neck provides a succinct pre-sentation of the relevant anatomy, directly applicable to day-to-day clinical scenarios.

The resource presented in this proposal therefore presents the reader with an easy access format to clinically applied anatomy perfect for revision purposes and quick reference to essential detail. This will make it ideal for the clinical arena, perfect for the pocket or on the wards, general practice office or emergency room. It will be ideal for the practitioner and student learning the anatomy of the head and neck and associated nerves and will directly relate to how the patient presents to the physician, surgeon, or allied health professional with nerve-related pathology.

In addition, it will also have labeled and unlabeled images of the nerves of the head and neck enabling the reader both to self-test and also confirm what the structures are in the professionally dissected and preserved specimens (prosections).

This final book in the series brings together material from the other three books (*Clinical Anatomy of the Cranial Nerves, Essential Clinical Anatomy of the Nervous System*, and *Essential Clinically Applied Anatomy of the Peripheral Nervous System in the Limbs*) in a summary format. It also introduces some new fields too but helps the aider have a greater understanding of head and neck anatomy, and how it links into the nervous syetm of the rest of the human body, and appreciate how busy the head and neck is anatomically.

Paul M. Rea, MBChB, MSc, PhD, DipFMS, MIMI, RMIP, FHEA, FRSA
Laboratory of Human Anatomy, School of Life Sciences,
College of Medical, Veterinary and Life Sciences,
University of Glasgow, Glasgow, United Kingdom

ACKNOWLEDGMENTS

There are several people whom I would like to thank in making this book possible.

First I would like to express my gratitude to Elsevier for having the time, patience, and faith in me while putting this together. They have really been the backbone to helping me realize my dream in publishing this.

I would like to dedicate this book to my mother Nancy, father Paul, and the dearest brother Jaimie. Thank you for being there and supporting me throughout everything—I am so proud of you all!

I would also like to extend a very special note of thanks to David Kennedy for hearing every update, word count by word count, as to the progress of this book, and for being there in so many ways! Thank you so much!

Also, thank you to a dear friend who has gone but not been forgotten—Mark Peters.

Thank you also to Elaine Jamieson, Richard Locke, and Leana Zaccarini for all the years of very special friendship.

Finally, thank you to a dear colleague, friend, and amazing mentor who has supported me from when I first started my career as an anatomist through to where I am today—Dr John Shaw-Dunn.

CHAPTER 1

Overview of the Nervous System

1.1 OVERVIEW OF THE NERVOUS SYSTEM

Broadly speaking, the nervous system is divided into two components—central and peripheral. The central nervous system (CNS) comprises the brain as well as the spinal cord. The peripheral nervous system (PNS) comprises all of the nerves—cranial, spinal, and peripheral nerves, including the sensory and motor nerve endings of these nerves.

1.2 DIVISIONS OF THE NERVOUS SYSTEM

1.2.1 Central Nervous System

The CNS is comprised of the brain as well as the spinal cord. The purpose of the CNS is to integrate all the body functions from the information it receives. Within the PNS, there are many, many nerves (group of nerve fibers together), however, the CNS does not contain nerves. Within the CNS, a group of nerve fibers traveling together is called a pathway, or tract. If it links the left and right hand sides it is referred to as a commissure.

1.2.1.1 Neurons

Within the CNS, there are many, many millions of nerve cells called neurons. Neurons are cells which are electrically excitable and transmit information from one neuron to another by chemical and electrical signals. There are three very broad classifications of neurons—sensory (which process information on light, touch, and sound to name some of the modalities), motor (supplying muscles), and interneurons (which interconnect neurons via a network).

Typically a neuron comprises some basic features, however there are a variety of specializations that some have dependent on the location within the nervous system. In general, a neuron has a cell body. Here, the nucleus—the powerhouse—of the neuron lies with its

Essential Clinically Applied Anatomy of the Peripheral Nervous System in the Head and Neck.
DOI: http://dx.doi.org/10.1016/B978-0-12-803633-4.00001-6

cytoplasm. At this point, numerous fine fibers enter called dendrites. These processes receive information from adjacent neurons keeping it up-to-date with the surrounding environment. This way the amount of information that a single neuron receives is significantly increased. From a neuron, there is a long single process of variable length called an axon. This conducts information away from the neuron. Some neurons however have no axons and the dendrites will conduct information to and from the neuron. In addition to this, a lipoprotein layer called the *myelin sheath* can surround the axon of a principal cell. This is not a continuous layer along the full length of the axon. Rather, there are interruptions called *nodes of Ranvier*. It is at this point where the voltage gated channels occur, and it is at that point where conduction occurs. Therefore, the purpose of the myelin sheath is to enable almost immediate conduction between one node of Ranvier and the next ensuring quick communication between neurons.

In relation to the size of neurons, this varies considerably. The smallest of neurons can be as small as 5 μm, with the largest for example motor neurons, can be as big as 135 μm. In addition, axonal length can vary considerably too. The shortest of these can be 100 μm, whereas a motor axon supplying the lower limb for example the toes, can be as long as 1 m.

In the PNS, neurons are found in *ganglia*, or in *laminae* (layers) or *nuclei* in the CNS.

Neurons communicate with each other at a point called a synapse. Most of these junctional points are chemical synapses where there is the release of a neurotransmitter which diffuses across the space between the two neurons. The other type of synapse is called an electrical synapse. This form is generally more common in the invertebrates, where there is close apposition of one cell membrane and the next that is at the pre- and postsynaptic membranes. Linking these two cells is a collection of tubules called *connexons*. The transmission of impulses occurs in both directions and very quickly. This is because there is no delay in the neurotransmitter having to be activated and released across the synapse. Instead, the flow of communication depends on the membrane potentials of the adjacent cells (Table 1.1).

Table 1.1 This Summarizes the Main Cellular Components of Nervous Tissue, Including the Role(s) for These	
Anatomical feature of neuron	**Function**
Soma (cell body)	Protein synthesis (abundance of Nissl substance/body) Location of neurofilaments (Maintenance of neuron and structural support) "Powerhouse" of neuron Location of nucleus, nucleolus and Nissl body (location of rough endoplasmic reticulum (and ribosomes))
Dendrites	Cellular extensions Majority of input to neuron arrives here via dendrites (via the dendritic spine)
Axon	Transmission of the electrical impulse (action potential) away from the neuronal cell body (soma) To allow for communication with nearby neurons
Axon terminal	Dilated terminal region of the axon Release of neurotransmitter (from the vesicles) into the synaptic cleft to communicate with the dendrite of the next neuron it targets
Axon hillock	The region close to the soma where the axon originates from Location of the voltage gated sodium channels Most excitable part of the neuron May receive information into this point too
Myelin sheath	Propagation of electrical impulses along the axon Increased electrical resistance No voltage gated channels
Node of Ranvier	Location of voltage gated ion channels Location of ion exchangers (e.g. Na^+/K^+ and Na^+/Ca^{2+} Aids rapid propagation of electrical impulses along the axon
Synapse	Structure capable of transmitting chemical or electrical signals Position of pre- and postsynaptic area for communication between two neurons (e.g. axon terminal of one neuron and dendrite of adjacent neuron)

1.2.1.2 Neuroglia

Neuroglia, or glia, are the supportive cells for neurons. Their main purpose is not in relation to the transmission of nerve impulses. Rather, they are involved in providing nutrient support, maintenance of homeostasis, and the production of the myelin sheath. There are two broad classifications—microglia and macroglia.

The microglia have a defence role as a phagocytic cell. They are found throughout the brain and spinal cord, and can alter their shape, especially when they engulf particulate material. They are thus serving a protective role for the nervous system.

Macroglia are subdivided into seven different types, again with each having a special role.

1. *Astrocytes*
   These cells fill in the spaces between neurons and provide for structural integrity. They also have processes which join to the capillary
   blood vessels. These are known as *perivascular end feet*. Therefore,
   with their close apposition to blood vessels, they are also thought to
   be responsible for metabolite exchange between the neurons and the
   vasculature. They are found in the CNS.
2. *Ependymal cells*
   There are three types of ependymal cells—ependymocytes, tanycytes, and choroidal epithelial cells. The ependymocytes allow for
   the free movement of molecules between the cerebrospinal fluid
   (CSF) and the neurons. Tanycytes are generally found in the third
   ventricle and can be involved in responding to changing hormonal
   levels of the blood derived hormones in the CSF. Choroidal epithelial cells are the cells which control the chemical composition of the
   CSF. They are found in the CNS.
3. *Oligodendrocytes*
   These cells are responsible for the production of myelin sheaths.
   They are found in the CNS.
4. *Schwann cells*
   Like oligodendrocytes, Schwann cells are responsible for the production of the myelin sheath, but in the PNS. They also have an
   additional role in phagocytosis of any debris; therefore help to clean
   the surrounding environment.
5. *Satellite cells*
   These cells surround those neurons of the autonomic system and
   also the sensory system. They maintain a stable chemical balance of
   the surrounding environment to the neurons. They are therefore
   found in the PNS.
6. *Radial glia*
   Radial glial cells act as scaffolding onto which new neurons migrate
   to. They are found in the CNS.
7. *Enteric glia*
   These cells are found within the gastrointestinal tract and aid digestion and maintenance of homeostasis. They are by their very nature
   found in the PNS.

1.2.1.3 Gray and White Matter

In the CNS, there are two clear differences between the structural components. It is divided by its appearance of either gray or white matter. Within the gray matter there are cell bodies and dendrites of efferent neurons, glial cells (supportive), fibers of afferent neurons and interneurons. The white matter on the other hand primarily consists of myelinated axons and the supportive glial cells. The purpose of the white matter is to allow for communication from one part of the cerebrum to the other, and also to communicate to other brain areas and carry impulses through the spinal cord.

1.2.1.4 Brain

The brain is comprised of three swellings which form during development—the forebrain (prosencephalon), midbrain (mesencephalon), and hindbrain (rhombencephalon). During development in mammals, the forebrain continues to grow, whereas in other vertebrates for example amphibians and fish, the three divisions remain in proportion to each other during growth.

The brain can also be subdivided into the following:

a. *Telencephalon* (cerebral hemispheres) + *Diencephalon* (thalamus and hypothalamus) = *FOREBRAIN*
b. *Mesencephalon* = *MIDBRAIN*
c. *Metencephalon* (pons, cerebellum, and the trigeminal, abducent, facial, and vestibulocochlear nerves) + *Myelencephalon* (medulla oblongata) = *HINDBRAIN*

Surrounding the core of the forebrain that is the diencephalon are the two large cerebral hemispheres (left and right), which constitute the cerebrum. The cerebrum is composed of three regions:

1. *Cerebral cortex*
   The cerebral cortex is the gray matter of the cerebrum. It is comprised of three parts based on its functions—motor, sensory, and association areas. The motor area is present in both cerebral cortices. Each one controls the opposite side of the body that is the left motor area controls the right side of the body, and vice versa. There are two broad regions—a primary motor area responsible for execution of voluntary movements, and supplementary area involved in selection of voluntary movements.

The sensory area receives information from the opposite side of the body ie, the right cerebral cortex receives sensory information from the left side of the body. In essence it deals with auditory information (via the primary auditory cortex), visual information (via the primary visual cortex), and sensory information (via the primary somatosensory cortex).

The association areas allow us to understand the external environment. All of the cerebral cortex is subdivided into lobes of the brain. These are:

a. Frontal lobes

Broadly speaking the frontal lobe deals with "executive" functions and our long-term memory. It also is the site of our primary motor cortex, toward its posterior part.

b. Parietal lobes

The parietal lobes are responsible for integration of sensory functions. It is the site of our primary somatosensory cortex.

c. Temporal lobes

The temporal lobes integrate information related to hearing, and therefore, is the site of our primary auditory cortex.

d. Occipital lobes

The occipital lobes integrate our visual information and function as the primary visual cortex.

2. *Basal ganglia*

The basal ganglia are three sets of nuclei—the *globus pallidus*, *striatum*, and *subthalamic nucleus*. These nuclei are found at the lower end of the forebrain and are responsible for voluntary movement, development of our habits, eye movements, and our emotional and cognitive functions.

3. *Limbic system*

The limbic system is comprised of a variety of structures on either side of the thalamus. It serves a variety of functions including long-term memory, processing of the special sense of smell (olfaction), behavior, and our emotions.

1.2.1.4.1 Thalamus

The thalamus is like a junction point of information. It is a relay point for all sensory information (apart from that related to smell). It also functions in the regulation of our wakened state, or sleep. In addition, it provides a connection point for motor information on its way to the cerebellum.

1.2.1.4.2 Hypothalamus

The hypothalamus, as its name suggests, is located below the thalamus. It secretes hormones influencing the pituitary gland, and in turn, a wide variety of bodily functions. It regulates autonomic activity ranging from temperature control, hunger, and our circadian rhythm and thirst.

1.2.1.4.3 Midbrain

The midbrain, as its name suggests, is found between the hindbrain below and the cerebral cortices above. Comprised of the *cerebral peduncles*, *cerebral aqueduct*, and the *tegmentum*, it is involved in motor function, arousal state, temperature control, and visual and hearing pathways.

1.2.1.4.4 Hindbrain

The lowest part of the brain developmentally is the hindbrain and comprises the pons, medulla, and the cerebellum. These areas control movement, cardiorespiratory functions, and a variety of bodily functions like hearing and balance, facial movement, swallowing, and bladder control. Therefore brainstem death that is death of these regions, is incompatible with life.

1.3 SPINAL CORD

In the gray matter of the spinal cord, three broad categories exist: (1) sensory cells concerned with sensory and reflex arcs; (2) motor cells leaving by the ventral roots to supply skeletal muscle, and (3) motor cells leaving by the ventral roots to go on to supply the autonomic ganglia. On examining the spinal cord in cross section, there is a butterfly-shaped gray matter surrounded by white matter (discussed later). The cells found in the gray matter are composed of cell columns in the rostro-caudal axis. Here there are cell bodies, axons, and dendrites, both of the myelinated and unmyelinated types.

In each half of the spinal cord there are three funiculi: the dorsal funiculus (between the dorsal horn and the dorsal median septum), the lateral funiculus (located where the dorsal roots enter and the ventral roots exit), and the ventral funiculus (found between the ventral median fissure and the exit point of the ventral roots).

Based on detailed studies of neuronal soma size (revealed using the Nissl stain), Rexed (1952) proposed that the spinal gray matter is arranged in the dorso-ventral axis into laminae and designated them into 10 groupings of neurons identified as I—X (Fig. 1.1).

Lamina I contains the terminals of fine myelinated and unmyelinated dorsal root fibers that pass first through the zone of Lissauer (dorsolateral funiculus) and then enter lamina I mediating pain and temperature sensation (Christensen and Perl, 1970; Menétrey et al., 1977; Craig and Kniffki, 1985; Bester et al., 2000). The neurons here have been divided into small neurons and large marginal cells characterized by wide-ranging horizontal dendrites (Willis and Coggeshall, 1991). They then synapse on the posteromarginal nucleus. From here the axons of these cells pass to the opposite side and ascend as the lateral spinothalamic tract.

Lamina II is immediately below lamina I, referred to as the substantia gelatinosa. Neurons here modulate the activity of pain and temperature afferent fibers. This lamina has been subdivided into an outer

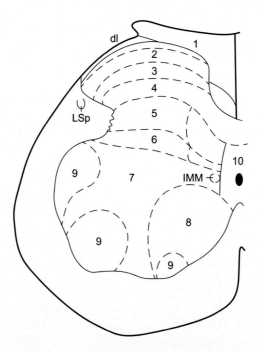

Figure 1.1 Cross section of the right side of the spinal cord showing the position of Rexed's laminae. 1—10 indicate the position of laminae I—X respectively. dl- dorsolateral funiculus; IMM—intermediomedial nucleus; LSp— lateral spinal nucleus.

(dorsal) lamina II (II_O) and an inner (ventral) lamina II (II_i) based on the morphology of these layers with stalked cells found in larger numbers in lamina II_O but stalked and islet cells were found throughout lamina II (Todd and Lewis, 1986). Lamina II is the region which receives an extensive unmyelinated primary afferent input, with very little from large myelinated primary afferents (except for distal parts of hair follicle afferents in some animals; Willis and Coggeshall, 1991). The axonal projections from here are wide and varied with some neurons projecting from the spinal cord (projection neurons), some passing to different laminae and some with axons confined to a lamina in the region of the dendritic tree of that cell for example intralaminar interneurons, local interneurons, and Golgi Type II cells (Todd, 1996).

Lamina III is distinguished from lamina II in that it has slightly larger cells, but with a neuropil similar to that of lamina II. The classical input to this lamina comes from hair follicles and other types of coarse primary afferent fibers which include Pacinian corpuscles and rapidly and slowly adapted fibers.

Lamina IV is a relatively thick layer that extends across the dorsal horn. Its medial border is the white matter of the dorsal column, and its lateral border is the ventral bend of laminae I–III. The neurons in this layer are of various sizes ranging from small to large and the afferent input here is from collaterals and from large primary afferent fibers (Willis and Coggeshall, 1991). Input also arises from the substantia gelatinosa (lamina II) and contributes to pain, temperature, and crude touch via the spinothalamic tract (Siegel and Sapru, 2006).

Lamina V extends as a thick band across the narrowest part of the dorsal horn. It occupies the zone often called the neck of the dorsal horn. It has a well-demarcated edge against the dorsal funiculus, but an indistinct lateral boundary against the white matter due to the many longitudinally oriented myelinated fibers coursing through this area. The cell types are very homogeneous in this area, with some being slightly larger than in lamina IV (Willis and Coggeshall, 1991). Again, like lamina IV, primary afferent input into this region is from large primary afferent collaterals as well as from receiving descending fibers from the corticospinal and rubrospinal tracts with axons also contributing to the spinothalamic tracts (Siegel and Sapru, 2006). In addition, in the thoracolumbar segments (T1–L2/3) the reticulated division of lamina V contains projections to sympathetic preganglionic neurons (Cabot et al., 1994).

Lamina VI is present only in the cervical and lumbar segments. Its medial segment receives joint and muscle spindle afferents, with the lateral segment receiving the rubrospinal and corticospinal pathways. The neurons here are involved in the integration of somatic motor processes.

Lamina VII present in the intermediate region of the spinal gray matter contains Clarke's nucleus extending from C8 to L2. This nucleus receives tendon and muscle afferents with the axons of Clarke's nucleus forming the dorsal spinocerebellar tract relaying information to the ipsilateral cerebellum (Snyder et al., 1978). Also within lamina VII are the sympathetic preganglionic neurons constituting the intermediolateral cell column in the thoracolumbar (T1–L2/3) and the parasympathetic neurons located in the lateral aspect of the sacral cord (S2–4). In addition Renshaw cells are located in lamina VII and are inhibitory interneurons which synapse on the alpha motor neurons and receive excitatory collaterals from the same neurons (Renshaw, 1946; Siegel and Sapru, 2006).

Lamina VIII and IX are found in the ventral gray matter of the spinal cord. Neurons here receive descending motor tracts from the cerebral cortex and the brainstem and have both alpha and gamma motor neurons here which innervate skeletal muscles (Afifi and Bergman, 2005). Somatotopic organization is present where those neurons innervating the extensor muscles are ventral to those innervating the flexors, and neurons innervating the axial musculature are medial to those innervating muscles in the distal extremities (Siegel and Sapru, 2006).

Lamina X is the gray matter surrounding the central canal and represents an important region for the convergence of somatic and visceral primary afferent input conveying nociceptive and mechanoreceptive information (Nahin et al., 1983; Honda, 1985; Honda and Lee, 1985; Honda and Perl, 1985). In addition lamina X in the lumbar region also contains preganglionic autonomic neurons as well as an important spinothalamic pathway (Ju et al., 1987a,b; Nicholas et al., 1999).

The *white matter* of the spinal cord contains the *ascending* and *descending* pathways. Some of these pathways ascend and descend to and from the brain, whereas others will connect to various levels within the spinal cord itself (Table 1.2).

Table 1.2 This Table Summarizes the Input to Each One of Rexed's Laminae, the Destination of Neurons in Each Layer, and an Overview of the Functions of Each of the Territories in the Spinal Cord

	Input	Destination	Information processed
I	Fine myelinated and unmyelinated dorsal root fibres	Lateral spinothalamic tract	Pain and temperature sensation
II	Unmyelinated primary afferent input	Projection neurons Variety of laminae Confined to the laminae of the dendritic tree of the neuron	Modulate the activity of pain and temperature afferent fibres
III	Hair follicles Pacinian corpuscles Rapidly and slowly adapted fibres	Deeper spinal laminae Posterior column nuclei Supraspinal relay centers	Mechanoreception Propriospinal pathways Pain, temperature and touch
IV	Collaterals Large primary afferent fibres Substantia-gelatinosa	Spinothalamic tract	Pain, temperature and crude touch
V	Large primary afferent Collaterals Corticospinal and rubrospinal tracts	Spinothalamic tract	Pain and temperature sensation
VI	Descending corticospinal and rubrospinal fibers (lateral segment) Joint and muscle spindle afferents (medial segment)	Innervation of limbs	Integration of somatic motor processes
VII	Tendon and muscle afferents	Spinocerebellar tract Sympathetic ganglia (in thoracic and upper lumbar regions) Parasympathetic fibers in S2-S4	Proprioception Visceral (autonomic) regulation
VIII	Descending motor tracts from the cerebral cortex and the brainstem	Motor neurons	Intrafusal muscle fibers
IX	Descending motor tracts from the cerebral cortex and the brainstem	Skeletal muscles	Posture and balance Distal muscle movement
X	Convergence of somatic and visceral primary afferent input Autonomic regulation (in lumbar region)		Nociceptive and mechanoreceptive information

1.4 PERIPHERAL NERVOUS SYSTEM

The nerves in the PNS transmit information from all parts of the body to and from the CNS. In total, there are 43 pairs of nerves in the PNS—12 cranial nerves and 31 spinal nerves.

The nerves of the PNS can either be myelinated (formed by the surrounding Schwann cells) or unmyelinated in nature. Whether they have this myelin or not, they do have the same general feature in that a nerve contains nerve fibers with axons of either afferent or efferent neurons. Therefore, the nerves of the PNS can be classified as belonging to either afferent (taking information to the CNS) or efferent (away from the CNS). With spinal nerves, they contain both afferent and efferent information, whereas some cranial nerves like the olfactory and optic nerves contain only afferent information (for smell and sight respectively).

Afferent information transmits impulses from receptors to the CNS. Their axons are found outwith the CNS, but then enter the CNS. Efferent information however transmits information from the CNS externally to, for example, glands and muscles. It is worthy to note that the efferent division is subclassified into what they ultimately supply. The further classification used of the efferent division is the somatic and autonomic nervous system (ANS). Simply put, the somatic nervous system innervates skeletal muscle, whereas the ANS innervates glands, neurons of the gastrointestinal tract, and cardiac and smooth muscles of glandular tissue.

1.4.1 Somatic Nervous System

The somatic nervous system consists of the cell bodies located in either the brainstem or the spinal cord. They have an extremely long course as they do not synapse after they leave the CNS until they are at their termination in skeletal muscle. They consist of large diameter fibers and are ensheathed with myelin. They are commonly referred to as motor neurons due to their termination in skeletal muscle. Within the muscle fibers, they release the neurotransmitter acetylcholine and are only excitatory that is result only in contraction of the muscle.

1.4.2 Cranial Nerves

There are 12 pairs of cranial nerves and these emerge either from the brain as fiber tracts (olfactory (I) and optic (II) nerves) or the brainstem (all other cranial nerves (III–XII)). The details of each of these nerves, their functions, and clinical applications will be dealt with in turn in subsequent chapters.

1.4.3 Spinal Nerves — Overview

There are 31 pairs of nerves that connect with the spinal cord as the spinal nerves. There are eight in the cervical region, twelve in the thoracic region, five in the lumbar region, five in the sacral region, and one coccygeal nerve.

1.4.4 Spinal Nerves

There are 31 pairs of spinal roots with their corresponding dorsal and ventral roots. There are eight *cervical*, twelve *thoracic*, five *lumbar*, five *sacral*, and one *coccygeal*. Within the spinal roots, there are dorsal and ventral roots. In the ventral roots, there are motor fibers, and it is those fibers that supply the skeletal muscle. Within the ventral roots of the thoracic, upper lumbar and some of the sacral levels autonomic fibers are also present. Within the dorsal roots, there are sensory fibers from the skin, subcutaneous and deep tissue, and frequently from the viscera too. The spinal nerve is formed by both the dorsal and ventral root and contains most of the fiber components found in that root. A major peripheral nerve therefore contains the sensory, motor, and autonomic fibers within it. Smaller branches however will vary in their composition. Therefore, a nerve to skin will lack motor fibers to the skeletal muscle, but it will contain sensory fibers and autonomic fibers to the vasculature and may also contain fibers supplying the autonomic innervation to the hair follicles.

The dorsal rami will transmit information from the muscles of the back and also the skin. The ventral rami innervate the rest of the trunk and also the limbs. The ventral rami that supply the thorax and abdomen remain relatively separate in their course. However in the cervical or lumbar and sacral regions, the ventral rami are found intertwined in what is called a plexus of nerves. It is from these plexuses that the peripheral nerves will then emerge.

A simple way to look at it is that when the ventral rami enter into this plexus, its individual components will contribute to several of the peripheral nerves. Indeed, each peripheral nerve therefore will contain fibers from one or more spinal nerves.

Each spinal nerve has a distribution called a *dermatome*. This is the area that is supplied by a single spinal nerve by its sensory fibers running in its dorsal root. However, sectioning of a single spinal nerve rarely will result in complete loss of sensation, or anesthesia. This is because other adjacent spinal nerves will also be carrying fibers from that site. Therefore the more likely scenario is a reduced level of sensation, or hypoesthesia.

There are many versions of *dermatome maps* which can be used clinically to demonstrate the site(s) of pathology of a patient with a suspected spinal nerve lesion. The most common one to be used clinically is the American Spinal Injury Association's (ASIA) worksheet produced as the International Standards for Neurological Classification of Spinal Cord Injury (ISNCSCI). This is discussed in considerable detail in Chapter 3, under the section on the supraclavicular nerve.

For motor fibers, carried in the ventral root, they tend to supply more than one muscle. Therefore, each muscle will receive innervation from more than one spinal nerve. Sectioning of a single spinal nerve will result in weakness of more than one muscle. Sectioning of a peripheral nerve will however result in paralysis of that single muscle. The group of muscles that a single spinal nerve supplies is called a *myotome*. Testing of the myotomes is something routinely carried out in the neurological examination of a patient, directed as to the signs and symptoms that the person presents with.

1.5 AUTONOMIC NERVOUS SYSTEM

The ANS can be thought of as that part of the nervous system supplying all other structures apart from skeletal muscle (supplied by the somatic nervous system). However, part of the ANS supplies the gastrointestinal system and is referred to as the enteric nervous system as the neurons are found supplying the glands and smooth muscle in the actual wall of the tract.

Within the ANS generally, it is composed of two neurons and a synapse. This is different to the single neuron of the somatic nervous system. The origin of the first neuron of the ANS is found in the CNS, with the first synapse occurring in an autonomic ganglion. This part is defined as the preganglionic fiber. After the synapse in the autonomic ganglion, the second fiber is referred to as the postganglionic fiber as it passes to the effector organ, in this case cardiac or smooth muscle, glands, or gastrointestinal neurons.

The ANS is subdivided into sympathetic and parasympathetic divisions based on physiological and anatomical differences. The sympathetic division arises from the thoracolumbar region from the first thoracic to the second lumbar level (T1−L2). The parasympathetic division arises from cranial and sacral origins. Specifically, the parasympathetic division arises from four of the cranial nerves—the oculomotor (III), facial (VII), glossopharyngeal (IX), and vagus (X) nerves. It also arises from the sacral plexus at the levels of the second to fourth sacral segments (S2−4).

1.5.1 Sympathetic Nervous System

The sympathetic nervous system arises from the thoracolumbar region of the spinal cord. Most of the sympathetic ganglia lie in close proximity to the spinal cord forming two chains on either side of the body. These are referred to as the *sympathetic trunks*. However, some ganglia lie a little further away from the spinal cord and are referred to as *collateral ganglia*. These are found close to the arteries in the abdomen with the same names that is coeliac, superior mesenteric, and inferior mesenteric ganglia. They tend to lie closer to the organs that they supply.

Although the sympathetic nervous system arises specifically at the thoracolumbar region that is from the first thoracic to either the second or third lumbar vertebral levels, the sympathetic trunk extends from the neck to the sacrum. This is because some of the preganglionic fibers arising from the thoracolumbar region travel either up or down several vertebral segments before forming their synapses with the respective postganglionic neurons. Within the neck, the cervical ganglia are referred to as the superior and middle cervical or stellate ganglia.

This allows the sympathetic nervous system to act as a single unit, but with small areas also able to act independently. This contrasts with the parasympathetic nervous system which tends to act independently. This arrangement is ideal to be involved in fine regulation of the activities of the organs or territories that they supply.

The sympathetic nervous system is responsible for the body's "fight or flight" reaction. Therefore, through its innervation of the adrenal medulla, which releases adrenaline (epinephrine) as its major secretion (80%; with the other 20% being noradrenaline (norepinephrine), it helps to protect the body in times of threat to it. It would therefore be involved in functions like dilating the pupil, increasing heart rate and contractility, relaxation of the bronchial muscle and reduction in secretion of the bronchial glands, and reduction of gut motility. This allows blood to be diverted to those areas in need if the body needs to "fight or flight." The adrenal medulla is a bit unusual in its innervation by the sympathetic nervous system, as the postganglionic side of the adrenal medulla never develops axons. Instead the preganglionic fibers terminating in the adrenal medulla result in the secretion from it of epinephrine/norepinephrine and is viewed as an endocrine gland as its secretions pass into the bloodstream.

1.5.2 Parasympathetic Nervous System

The parasympathetic nervous system is described as originating in the craniosacral region that is from the brainstem and also the sacral plexus. Specifically, the parasympathetic nervous system, from the cranial side, concerns four of the cranial nerves, which will be dealt with later (specifically the oculomotor, facial, glossopharyngeal and vagus nerves). Specifically, the nuclei related to these are the Edinger-Westphal nucleus for the oculomotor nerve, superior salivatory and lacrimal nuclei for the facial nerve, inferior salivatory nucleus for the glossopharyngeal nerve, and the dorsal nucleus of the vagus nerve, as well as the nucleus ambiguus for the vagus nerve. This is where the preganglionic fibers are found for the parasympathetic nervous system. In addition to this, the sacral parasympathetic nucleus arising from the second, third, and fourth sacral segments are also involved.

The parasympathetic nervous system is opposite in its functions generally to the sympathetic nervous system. It can informally be referred to as the part of the nervous system responsible for "rest and digest". This part of the nervous system is responsible for the internal functions when you are sitting resting and relaxing. Therefore, it would constrict the pupil, slow heart rate and contractility, contract bronchial musculature and stimulate bronchial secretions, and enhance gut motility for digestion to effectively occur.

The main neurotransmitter in both the sympathetic and parasympathetic nervous systems at the preganglionic fiber, as it contacts the postganglionic fiber is acetylcholine. The same is also true of the postganglionic fiber as it contacts the effector organ generally. Therefore, where acetylcholine is secreted, it is referred to as *cholinergic*. However, in the sympathetic nervous system, the major neurotransmitter between the postganglionic fiber and the effector organ tends to be noradrenaline (norepinephrine). It also tends to be the case that this is not an exclusive relationship as to what is secreted, and at what site it is secreted. In addition to this, co-transmitters tend to also be present for example ATP, dopamine, and other neuropeptides.

1.6 FUNCTIONAL DIVISION OF THE NERVOUS SYSTEM

The following diagram summarizes the functional divisions of the nervous system. Central to this, is the brain and spinal cord (CNS). Information from the periphery arriving into the CNS is referred to as afferent. Information exiting the CNS is referred to as efferent. There are two afferent inputs which enter the CNS—somatic and visceral.

Tip!

The easy way to remember what information arrives into and leaves the CNS is rather easy.

*A*FFERENT—*A*RRIVES into the CNS. *A* for *AFFERENT*, *A* for *ARRIVES!*
*E*FFERENT—*E*XITS the CNS. *E* for *EFFERENT*, *E* for *EXIT!*

1.6.1 Somatic Nervous System

The somatic nervous system consists of the cell bodies located in either the brainstem or the spinal cord. They have an extremely long course as they do not synapse after they leave the CNS. These fibers synapse only when they terminate in the skeletal muscle. They consist of large diameter fibers and are ensheathed with myelin. They are commonly referred to as motor neurons due to their termination in skeletal muscle. Within the muscle fibers, they release the neurotransmitter acetylcholine and are only excitatory that is result only in contraction of the muscle.

1.6.2 Autonomic Nervous System

The visceral system, or ANS, can be thought of as that part of the nervous system supplying all other structures apart from skeletal muscle (supplied by the somatic nervous system). However, part of the ANS supplies the gastrointestinal system and is referred to as the enteric nervous system as the neurons are found supplying the glands and smooth muscle in the actual wall of the tract.

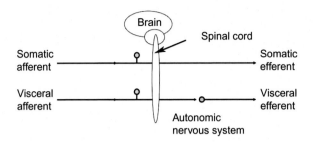

Figure 1.2 A diagram showing the functional divisions of the nervous system—somatic and visceral. Information arriving into the CNS is referred to as afferent, and information leaving it is efferent. Note the 2 neuron pathway in the autonomic (visceral) nervous system.

Within the ANS generally, it is composed of two neurons and a synapse (as shown in Figure 1.2). This is different from the single neuron of the somatic nervous system. The origin of the first neuron of the ANS is found in the CNS, with the first synapse occurring in an *autonomic ganglion*. This part is defined as the *preganglionic fiber*. After the synapse in the autonomic ganglion, the second fiber is referred to as the *postganglionic fiber* as it passes to the *effector* organ, in this case *cardiac* or *smooth muscle*, *glands*, or *gastrointestinal neurons*.

The ANS is subdivided into *sympathetic* and *parasympathetic* divisions based on physiological and anatomical differences. The *sympathetic* division arises from the *thoracolumbar* region from the first thoracic to the second lumbar level (T1−L2). The *parasympathetic* division arises from *cranial* and *sacral* origins. Specifically, the parasympathetic division arises from four of the cranial nerves—the *oculomotor* (III), *facial* (VII), *glossopharyngeal* (IX), and *vagus* (X) nerves. It also arises from the *sacral plexus* at the levels of the second to fourth sacral segments (S2−4).

1.6.2.1 Sympathetic and Parasympathetic Nervous System
The sympathetic nervous system is part of the nervous system that deals with *"fight or flight"* responses. The parasympathetic nervous system can be classified as part of the nervous system that controls *"rest and digest."* A summary table is given below comparing what functions each part of the ANS causes to a variety of areas around the body (Table 1.3).

Table 1.3 This Table Compares the Differences at Various Regions of the Body of the Sympathetic and Parasympathetic Nervous System

	Sympathetic	Parasympathetic
Heart	Increases heart rate	Reduces heart rate
	Increases contractility of atria and ventricles	Reduces contractility of atria and ventricles
	Increases conduction	Reduces conduction
Lungs	Relaxes bronchial muscle	Contracts bronchial muscle
	Reduced secretions (via $\alpha1$ receptors)	Stimulates secretions (via $\alpha1$ receptors)
Stomach and intestines	Reduced tone and motility	Increased tone and motility
	Contracts sphincters	Relaxes sphincters
	Inhibits secretions	Stimulates secretions
Pancreas	Inhibits exocrine secretion	Stimulates secretion
	Inhibits insulin secretion	Stimulates insulin secretion
Eyes	Contracts radial muscle (dilates pupil)	Contracts sphincter muscle (constricts pupil)
	Relaxes ciliary muscle (for far vision)	Contracts ciliary muscle (for near vision)
Nasal, lacrimal and salivary glands	No significant effect	Stimulation of serous and mucous secretions from the secretory cells
Skin	Contracts arrector pili muscles (hair to stand on end)	N/A
	Localised secretion of sweat glands	Generalised secretion of sweat glands
Urinary bladder	Relaxes wall	Contracts wall
	Contracts sphincter	Relaxes sphincter
Genital organs	May stimulate vasoconstriction, but uncertain and variable	May stimulate glands and smooth muscle; vascular dilatation
Adrenal gland	Stimulation of secretory cells to produce epinephrine	No effect
Arterioles	Variable	Dilates coronary and salivary gland arterioles (via $\alpha1,2$ receptors)

The different divisions of the autonomic nervous system affect each territory in very contrasting ways.

REFERENCES

Afifi, A., Bergman, R., 2005. Functional Neuroanatomy, second ed. McGraw-Hill, USA.

Bester, H., Chapman, V., Besson, J.M., Bernard, J.F., 2000. Physiological properties of the lamina I spinoparabrachial neurons in the rat. J. Neurophysiol. 83, 2239−2259.

Cabot, J.B., Alessi, V., Carroll, J., Ligorio, M., 1994. Spinal cord lamina V and lamina VII interneuronal projections to sympathetic preganglionic neurons. J. Comp. Neurol. 347, 515−530.

Christensen, B.N., Perl, E.R., 1970. Spinal neurons specifically excited by noxious or thermal stimuli: marginal zone of the dorsal horn. J. Neurophysiol. 33, 293−307.

Craig, A.D., Kniffki, K.D., 1985. Spinothalamic lumbosacral lamina I cells responsive to skin and muscle stimulation in the cat. J. Physiol. 365, 197−221.

Honda, C.N., 1985. Visceral and somatic efferent convergence onto neurons near the central canal in the sacral spinal cord of the cat. J. Neurophysiol. 53, 1059−1078.

Honda, C.N., Perl, E.R., 1985. Functional and morphological features of neurons in the midline region of the caudal spinal cord of the cat. Brain Res. 340, 285−295.

Ju, G., Hökfelt, T., Brodin, E., Fahrenkrug, J., Fischer, J.A., Frey, P., et al., 1987a. Primary sensory neurons of the rat showing calcitonin gene-related peptide immunoreactivity and their relation to substance P, somatostatin, galanin, vasoactive intestinal polypeptide and cholecystokinin immunoreactive ganglion cells. Cell Tissue Res. 247, 417−431.

Ju, G., Melander, T., Ceccatelli, S., Hökfelt, T., Frey, P., 1987b. Immunohistochemical evidence for a spinothalamic pathway co-containing cholecystokinin and galanin like immunoreactivities in the rat. Neuroscience 20, 439−456.

Menétrey, D., Giesler Jr., G.J., Besson, J.M., 1977. An analysis of response properties of spinal dorsal horn neurons to non-noxious and noxious stimuli in the rat. Exp. Brain Res. 27, 15−33.

Nahin, R.L., Madsen, A.M., Giesler, G.J., 1983. Anatomical and physiological studies of the grey matter surrounding the spinal cord central canal. J. Comp. Neurol. 220, 321−335.

Nicholas, A.P., Zhang, X., Hökfelt, T., 1999. An immunohistochemical investigation of the opioid cell column in lamina X of the male rat lumbosacral spinal cord. Neurosci. Lett. 270, 9−12.

Renshaw, B., 1946. Central effects of centripetal impulses in axons of spinal ventral roots. J. Neurophysiol. 9, 191−204.

Rexed, B., 1952. The cytoarchitectonic organisation of the spinal cord in the cat. J. Comp. Neurol. 96, 414−495.

Siegel, A., Sapru, H.N., 2006. Essential Neuroscience. Lippincott Williams and Wilkins, Baltimore.

Snyder, R.L., Faull, R.L., Mehler, W.R., 1978. A comparative study of the neurons of origin of the spinocerebellar afferents in the rat, cat and squirrel monkey based on the retrograde transport of horseradish peroxidase. J. Comp. Neurol. 15, 833−852.

Todd, A.J., 1996. GABA and glycine in synaptic glomeruli of the rat spinal dorsal horn. Eur. J. Neurosci. 8, 2492−2498.

Todd, A.J., Lewis, S.G., 1986. The morphology of Golgi- stained neurons in lamina II of the rat spinal cord. J. Anat. 149, 113−119.

Willis, W.D., Coggeshall, R.E., 1991. Sensory Mechanisms of the Spinal Cord, second ed. Plenum Press, New York.

CHAPTER 2

Head

2.1 INTRODUCTION

The head and neck is perhaps one of the most complicated parts of the human body when it comes to learning the anatomy of it. Many structures are compacted into such a small space. Even when the superficial structures are found within the dissecting room, the origins of such structures are actually found much deeper and take some time to understand the full pathway of these parts.

The head is defined as the uppermost part of the body and connects the trunk via the neck. The head contains the special sensory receptors and related organs, as well as major blood vessels, nerves, and the brain, essential for life and higher order processes.

The head comprises the brain as well as its protective coverings called the meninges. The head also includes the face as well as the ears. The face provides our unique identity and houses the openings of the nose and mouth, with its related glands and muscles for mastication (chewing). It also houses our eyes in a perfect position for our binocular vision.

The skull is a logical consequence of the process of cephalization—an evolutionary trend in which multicellular organisms have developed an elongated, cylindrical body with a leading (front) and trailing (back) end. There are clear advantages in grouping sense organs at the front, and also in making the front end the site of ingestion (food, air, water); to cope with these sense organs, the nervous system enlarges at the front (ie, the brain is developed) and there is a simultaneous requirement to protect these valuable acquisitions—hence the skull.

Essential Clinically Applied Anatomy of the Peripheral Nervous System in the Head and Neck.
DOI: http://dx.doi.org/10.1016/B978-0-12-803633-4.00002-8

The skull is composed of a number of individual bones but is best appreciated initially as a totality. It can conveniently be divided into three parts:

1. The *cranium* which is in turn divided into:
   a. the *neurocranium*, essentially a rounded container for the brain and special senses, and
   b. the *viscerocranium*, the irregularly shaped part at the front which provides a skeletal framework for the nasal and oral cavities.
2. The *mandible:*

   The neurocranium consists of the rounded vault of the skull and a rather irregular floor, the base of the skull. The bones of the vault are large and easily identified. It comprises the frontal bone, two parietal bones and the occipital bone, and the two temporal bones at the side. The bones meet (*articulate*) at wavy lines called *sutures*. These are the coronal, sagittal, and lambdoid sutures. These are immobile, fibrous joints. The newborn baby has gaps between many of the bones called *fontanelles* (the "soft spots") where the bony plates have yet to grow and meet. The bones of the vault actually develop within a membranous sheet and it is this sheet which fills the interval. The anterior fontanelle is a particularly large, diamond-shaped space which does not fully close until *two years after birth*. If the baby is dehydrated, the surface of the head appears to sink in at this point.

The frontal, parietal, and much of the occipital bones are covered in life by the scalp, which can be moved backwards and forwards by the *occipito-frontalis muscle*. The scalp has a rich blood and nerve supply and is such a dense, taut structure that, when wounding occurs, the cut ends of blood vessels tend to pull apart. Thus, scalp wounds bleed profusely and injury often appears worse than it really is.

The sides of the neurocranium are chiefly composed of the *temporal bone*. This is an extremely complex bone and it is chiefly the flattened squamous part which can be seen from the outside of the skull (a projection from this makes a contribution to the *zygomatic arch* (cheek bone). In life, the *temporal bone* gives origin to (and is covered by) the temporalis muscle.

The *occipital bone* can be followed downwards to form part of the base of the skull. This ventral part of the bone has two important features (1) the foramen magnum ("big hole") through which the spinal cord and brain are continuous and (2) the rounded occipital condyles, by which the skull articulates with the first cervical vertebra—the atlas.

The floor of the interior of the skull of the neurocranium has three levels or "steps," the highest at the front and the lowest at the back. These are the floors of the three cranial fossae (anterior, middle, and posterior). The anterior cranial fossa houses the frontal lobes of the brain and is chiefly formed by the frontal bone, especially the orbital plate (which, as its name implies, forms the roof of the orbit). Behind this is the lesser wing of the sphenoid and, in the midline, the cribriform (sieve-like) plate of the ethmoid bone; this is perforated by many holes which transmit olfactory nerves from the nasal cavity which lies beneath it.

The floor of the middle cranial fossa is chiefly formed by the sphenoid bone (which lies in the central part of the fossa) and the temporal bone (which lies at the side). For the sphenoid bone, it comprises the greater wing and the body; the latter has a concave region where the pituitary gland is situated called—rather fancifully—the *sella turcica* ("Turkish saddle"). There are a number of foramina in (or at the edges of) the sphenoid bone which transmit cranial nerves. On each side are:

1. the *optic canal* (optic nerve and ophthalmic artery)
2. the *superior orbital fissure* (ophthalmic veins, nerves to extraocular muscles, ophthalmic division of trigeminal nerve)
3. the *foramen rotundum* (maxillary division of trigeminal nerve)
4. the *foramen ovale* (mandibular division of trigeminal nerve)

The floor of the posterior cranial fossa is chiefly formed by the occipital bone. Also present are:

1. the *jugular foramen* (through which the glossopharyngeal (IX), vagus (X), and accessory (XI) nerves leave the skull, together with the sigmoid venous sinus which forms the internal jugular vein) and
2. the *hypoglossal canal* (through which the hypoglossal nerve leaves the skull).

2.2 BONES OF THE SKULL

The skull comprises many bones which are fixed to each other by immobile joints. However, the mandible, as mentioned is considered separate, and is also freely mobile, connected to the skull with a synovial joint at the temporomandibular joint (TMJ).

2.3 BRAIN

The human brain has been defined as an incredibly complex organ, and we are still gaining new evidence in relation to its functions, and what happens pathologically. The "average" human adult brain comprises approximately 2% of body weight, and its weight can range from 1.2 to 1.4 kg. The brain is an incredibly demanding organ and is essential for life and function of our body. As such, it consumes an enormous amount of the circulating blood volume. Approximately one-sixth of all cardiac output passes through the brain at any one time, and it uses about one-fifth of all the oxygen in the body when we are at rest.

The brain is our most complex organ and controls and regulates our body, responds to stress and threat, and controls higher cognitive functions. It maintains body temperature, allows us to interpret the special senses, and to socially interact. It ensures the body works optimally in the environment we are in both protecting and nurturing the human body.

2.3.1 Divisions of the Brain

All vertebrates share the same basic structure to them. The basis of the embryological development is from the neural tube, or the precursor of the central nervous system. At the front, or upper end, of that tube-like structure, there are three swellings. These three swellings then become the forebrain, midbrain, and hindbrain. The human brain is broadly divided into these three main regions—the forebrain, midbrain, and the hindbrain. In mammals, the first part of this neural tube—the forebrain—becomes considerably larger, with the hindbrain remaining rather small in comparison. As previously discussed, the other way to classify the brain based on its components is as follows:

1. *Telencephalon* (cerebral hemispheres) + *Diencephalon* (thalamus and hypothalamus) = *Forebrain.*
2. *Mesencephalon* = *Midbrain.*

3. *Metencephalon* (pons, cerebellum, and the trigeminal, abducens, facial, and vestibulocochlear nerves) + *Myelencephalon* (medulla oblongata plus the glossopharyngeal, vagus, accessory, and hypoglossal nerve nuclei) = *Hindbrain.*

Surrounding the core of the forebrain, for example, the diencephalon is the two large cerebral hemispheres (left and right), which constitutes the cerebrum. The cerebrum is composed of three regions:

1. *Cerebral cortex*

   The cerebral cortex is the grey matter of the cerebrum. It comprises three parts based on its functions—motor, sensory, and association areas. The motor area is present in both cerebral cortices. Each one controls the opposite side of the body, that is, the left motor area controls the right side of the body, and vice versa. There are two broad regions—a primary motor area responsible for execution of voluntary movements and supplementary areas involved in selection of voluntary movements.

   The sensory area receives information from the opposite side of the body, that is, the right cerebral cortex receives sensory information from the left side of the body and vice versa. In essence, it deals with auditory information (via the primary auditory cortex), visual information (via the primary visual cortex), and sensory information (via the primary somatosensory cortex).

   The association areas allow us to understand the external environment. All of the cerebral cortex is subdivided into lobes of the brain. These are:

   a. Frontal lobes

      Broadly speaking the frontal lobe deals with "executive" functions and our long-term memory. It also is the site of our primary motor cortex, towards its posterior part.

   b. Parietal lobes

      The parietal lobes are responsible for integration of sensory functions. It is the site of our primary somatosensory cortex.

   c. Temporal lobes

      The temporal lobes integrate information related to hearing, and therefore, is the site of our primary auditory cortex.

   d. Occipital lobes

      The occipital lobes integrate our visual information and functions as the primary visual cortex.

2. *Basal ganglia*

The basal ganglia are three sets of nuclei—the *globus pallidus*, *striatum*, and *subthalamic nucleus*. These nuclei are found at the lower end of the forebrain and are responsible for voluntary movement, development of our habits, eye movements, and our emotional and cognitive functions.

3. *Limbic system*

The limbic system comprises a variety of structures on either side of the thalamus. It serves a variety of functions including long-term memory, processing of the special sense of smell (olfaction), behavior, and our emotions.

2.3.2 Thalamus

The thalamus is like a junction point of information. It is a relay point for all sensory information (apart from that related to smell). It also functions in the regulation of our wakened state or sleep. In addition, it provides a connection point for motor information on its way to the cerebellum.

2.3.3 Hypothalamus

The hypothalamus, as its name suggests, is located below the thalamus. It secretes hormones influencing the pituitary gland, and, in turn, a wide variety of bodily functions. It regulates autonomic activity ranging from temperature control, hunger, and our circadian rhythm and thirst.

2.3.4 Midbrain

The midbrain, as its name suggests, is found between the hindbrain below and the cerebral cortices above. Composed of the *cerebral peduncles*, *cerebral aqueduct*, and the *tegmentum*, it is involved in motor function, arousal state, temperature control, and visual and hearing pathways.

2.3.5 Hindbrain

The lowest part of the brain developmentally is the hindbrain and comprises the pons, medulla, and the cerebellum. These areas control movement, cardiorespiratory functions, and a variety of bodily functions like hearing and balance, facial movement, swallowing, and bladder control. Therefore, brainstem death, that is, death of these regions, is incompatible with life.

Table 2.1 provides a broad overview of each of these regions before they will be discussed in more detail later.

Table 2.1 Functions Associated With the Forebrain, Midbrain, and Hindbrain

Region of forebrain	Functions / information processed
	Forebrain
Cerebral hemispheres (TELENCEPHALON)	The cerebral hemispheres process information related to "higher order" functions:

	Somatosensory information from the opposite side of the body **Motor control** of the opposite side of the body
	    - Selection and execution of voluntary movements
	    - Movement in space
	Planning and organization
	Memory
	Thought
	Emotions
	Problem solving
	Consciousness
	Attention
	Intelligence
	Language comprehension
	Processing of information related to **vision** and sound
	Speech production and articulation

Thalamus (DIENCEPHALON)	The thalamus can be thought of as a relay and modulating center:
	Modulation of **motor functions** (via the ventrolateral and ventroanterior regions of the thalamus) **Somatosensory relay** **Olfactory relay** (via the amygdala and prepyriform cortex projections to the mediodorsal thalamus, and on to the frontal lobe, or simply passing through the thalamus without synapse) **Visual relay** (via the lateral geniculate nucleus)
	Taste relay (via the ventral posteromedial nucleus) **Auditory relay** (via the medial geniculate nucleus in the thalamus) **Vestibular relay** (via the ventral posterolateral nuclei) **Thermal relay** (via the ventral posterolateral nuclei) **Sleep-wake cycle** **Arousal** **Consciousness** **Memory (recollection, familiarization, spatial)**
Hypothalamus (DIENCEPHALON)	The hypothalamus has many functions including circadian rhythm, maintenance of body temperature within a narrow effective range, adrenocortical regulation. Broadly speaking , the hypothalamus may be sub-divided into two main territories – lateral and medial.

	Lateral hypothalamus
	    - **Thirst center, hunger center, predation, reward, motivation**
	    - stimulation of **parasympathetic** outflow
	    - **emotions** and **behavior**
	    - descending **modulation** of spinal neuronal activity

	Medial hypothalamus
	    - control of the **pituitary gland**
	    - stimulation of **sympathetic** outflow
	    - motivational reactions to **noxious** stimuli
	    - **emotions** and **behavior**

Region of forebrain	Functions / information processed
Midbrain	
Mesencephalon	**Auditory pathway** **Vision** (the oculomotor and trochlear nerves arise from this point) **Autonomic functions** **Emotional and affective processes** **Modulation of pain** **Somatomotor function** **Autonomic** and **visceral** functions **Ascending sensory pathways** (spinothalamic, medial lemniscus, trigeminothalamic, auditory)
Hindbrain	
Pons (METENCEPHALON)	Relay center between the cerebral hemispheres and the cerebellum (e.g. medial longitudinal fasciculus and medial lemniscus, spinothalamic, trigeminothalamic, corticobulbar tract, corticospinal tract, rubrospinal tract, tectospinal tract) Location of abducent, facial, trigeminal, superior and lateral vestibular and superior olivatory nuclei, reticular formation and cerebellar peduncles The pons controls a variety of functions – **sleep, respiration** (via the **pontine respiratory group**), **swallowing, auditory processing, control of the bladder, control of equilibrium, facial expression and sensation**
Cerebellum (METENCEPHALON)	Coordination and regulation of motor control (**locomotion**) **Control of balance and posture** **Vestibular input** **Cognitive functions** (Goldman-Rakic, 1996; Schmahmann and Caplan, 2006) Connections with the hypothalamus for **autonomic** and **emotional functions** (Schmahmann and Caplan, 2006)
Trigeminal nerve (METENCEPHALON)	**Sensation from face; paranasal sinuses; nose and teeth** **Muscles of mastication**
Abducent nerve (METENCEPHALON)	Innervates the **lateral rectus** muscle
Facial nerve (METENCEPHALON)	**Muscles of facial expression, stylohyoid, stapedius, posterior belly of digastric** Parasympathetic innervation of the **submandibular** and **sublingual salivary glands, lacrimal gland** and the **nasal** and **palatal glands** Anterior two-thirds of the tongue (taste) and palate (**general visceral afferent**) **Concha of the auricle** (general sensation)
Vestibulocochlear nerve (METENCEPHALON)	**Balance** for the **vestibular component** **Hearing** for the **spiral (cochlear) component**
Medulla oblongata (MYELENCEPHALON)	**Cardio-respiratory center** **Reflex centers** (e.g. swallowing, vomiting, sneezing, coughing)**Vasomotor center**

2.4 CRANIAL NERVES

The peripheral nervous system is the part of the nervous system that comprises the cranial, spinal, and peripheral nerves, as well as their sensory and motor nerve endings. In other words, it is the part of the nervous system which comprises nerves and ganglia which lie out with the brain and spinal cord (which comprises the CNS). As this part of the nervous system primarily lies out with the skull and vertebral column, it is prone to damage from trauma and from toxins. Therefore, the PNS comprises the 12 pairs of cranial nerves and the 31 pairs of spinal nerves, that is, 43 pairs of nerves in total. The nerves of the PNS can be classified as belonging to either afferent (taking information to the CNS) or efferent (away from the CNS). With spinal nerves, they contain both afferent and efferent information, whereas some cranial nerves like the olfactory and optic nerves contain only afferent information (for smell and sight, respectively).

It is essential for those involved in clinical professions like medicine, surgery, dentistry, nursing, and allied health care workers in the fields of neurology, to have a comprehensive knowledge and understanding of the anatomy and clinical applications of the cranial nerves. A detailed and thorough knowledge of the course of the cranial nerves through the cranial cavity and onwards to their final structures they innervate is crucial in neurological diagnoses. As the cranial cavity is an enclosed space, where the cranial nerves originate from, or pass to, even relatively small lesions may be clinically localized by the effect that they have on the function of a nearby nerve.

There are 12 pairs of cranial nerves, that is, 12 on the right hand side and 12 on the left hand side. These arise from the brain and have either a longer or shorter course to their foramina and then exit. Some of these nerves serve only a single purpose, whereas many of the cranial nerves carry a mixture of fibers within them, for example, sensory or motor to visceral or somatic structures, or indeed to structures from the branchial arches. Each nerve therefore can have connections with several nuclei in its course. In saying that, a few generalizations can still be made.

1. *Sensory fibers*
   These types of fibers run from the periphery uninterrupted to the point within the brainstem that they synapse with a second-order neuron. However, they have the cell body outside the brainstem. These cell bodies are collected together to form ganglia, for example, like the *trigeminal* or *facial ganglia*. These ganglia of the cranial nerves correspond to the spinal ganglia of the spinal cord. However, the exception to this is that the trigeminal nerve.
2. *Motor fibers*
   These supply skeletal muscle and have their cell body within the brainstem and an uninterrupted axon which then goes on to innervate the muscle. Therefore, there is no peripheral synapse.

3. *Sympathetic fibers*

These types of fibers are found within the face and skull and are always postganglionic.

4. *Parasympathetic fibers*

The preganglionic fibers synapse about the cell body of a postganglionic fiber in a peripherally located ganglion. The postganglionic fiber may well actually pass to another cranial nerve for distribution to other structures, or territories. Many of these ganglia exist within the head and neck. There are also fibers which pass through these ganglia but do not synapse there, or have a cell body within the ganglion.

Cranial nerves arise from the brain, as distinct from spinal nerves which arise from the spinal cord. There are 12 pairs of cranial nerves. Some are purely motor (eg, hypoglossal (XII) which supplies the tongue muscles), some are purely sensory (eg, the optic (II) nerves which come from the retina), and some are mixed (eg, the trigeminal (V) nerve which is sensory to the face and scalp and motor to the muscles of mastication). We can consider:

1. *Where they are attached to the brain :*
   a. Forebrain-olfactory (I) and optic (II) nerves
   b. Midbrain-oculomotor (III) and trochlear (IV) nerves (The trochlear is unique in arising from the dorsal surface of the brain stem)
   c. Hindbrain-trigeminal (V) from the pons; abducens (VI), facial (VII), and vestibulo-cochlear (VIII) from the ponto-medullary junction; glossopharyngeal (IX), vagus (X), accessory (XI), and hypoglossal (XII) from the medulla
2. *How they are related to embryological development.*

Many of the structures in the head and neck arise from two quite distinct embryological sources. These sources are:

1. *Somites*

These are paired segmental blocks of tissue which run along the length of the embryo, rather like a series of building blocks. They give rise to many structures, including muscles. Motor cranial nerves which supply somite-derived muscles are called somatic efferent[1] nerves and consist of cranial nerves III, IV, and VI which supply the extraocular muscles (which move the eyes about) and cranial nerve XII to the muscles of the tongue.

---

[1]"Efferent" means "going away from." In the context of cranial or spinal nerves, efferent nerves are those going out from the CNS, that is, motor nerves. The opposite is "afferent" (=going towards) which would describe sensory nerves taking information to the central nervous system.

2. *Branchial arches*

During development, the embryo passes through a stage of having pharyngeal or branchial arches at the side of the neck—exactly as a fish has gill arches. These arches form a numbered series and, again, give rise to many adult structures, including muscles. Motor cranial nerves which supply them are termed branchial efferent[1] nerves (Table 2.2).

Arch		Nerve	Muscles
I	Mandibular arch	Trigeminal (V)	Mastication
II	Hyoid arch	Facial (VII)	Facial expression
III – VI		IX, X, XI	Laryngeal/Pharyngeal

Table 2.2 Cranial Nerves Associated With Each of the Branchial Arches, and the Muscles They Supply

Here, each cranial nerve will be dealt with briefly, in terms of the ganglion (ganglia) associated with it, the type of fibers found within, and the function of each one.

2.4.1 Olfactory Nerve

The first cranial nerve is the olfactory nerve. It is the shortest of all the cranial nerves and is one of only two nerves that do not join with the brainstem. It is the cranial nerve responsible for conduction of impulses related to the special sense of smell, that is, special sensory. Specifically, it is a special visceral afferent nerve.

The cell bodies of the olfactory receptor neurons are located in the olfactory organ. These are found in the upper part of the nasal cavity, nasal septum, and on the inner aspect of the superior nasal concha. The olfactory receptor nerves have two surfaces, basal and apical, and these receive information from the odors which then dissolve in the mucous fluid to allow for electrical transmission of those impulses.

When the odor dissolves in the mucous, it is then detected by the olfactory nerves. Each olfactory nerve has two components: an apical and basal division. It is the cilia on the apical portion that detects the dissolved "smell." This then passes to the basal portion which constitutes the main processes of the olfactory nerve. The olfactory nerves at that point then enter the cranial cavity via the cribriform plate of the ethmoid bone. The special sense of smell is then transmitted towards the olfactory bulb where the cells then synapse. The transmission of the impulses carries posteriorly towards the brain via the olfactory tract. Specifically the information passes to the piriform cortex of the anterior temporal lobe, anterior olfactory nucleus, amygdala, and entorhinal cortex.

In summary, the odor dissolves in the mucous. The cilia on the apical surface of the olfactory receptor cells detect this. The basal surface of the olfactory receptor cells then becomes activated, which then, in turn, activates the olfactory bulb, sending the impulses to the various brain regions for olfaction.

The fiber types present in the olfactory nerve are:

Special Sensory
- Smell from the nasal septum, superior concha, and the roof of the nasal cavity. It is a special visceral afferent nerve.

The branches of the olfactory nerve are as follows:

Within the nasal cavity, there are two types of fibers—those of the trigeminal nerve (see Trigeminal Nerve chapter) which responds to irritating substances and temperature and those of the olfactory nerve for olfaction.

The olfactory receptor neurons have two parts:

a. Olfactory neurons in the olfactory epithelium have central processes which pass to the olfactory bulb where they synapse.
b. The processes from the synapse pass via the olfactory tracts to the brain regions for olfaction.

Clinical Examination
As with all clinical examinations, introduce yourself to the patient, stating who you are and your purpose for meeting with the patient, especially if you are a student. Always take a comprehensive history before examining the patient. This will guide you to the most appropriate examinations that need to be conducted. Ensure you explain everything to the patient prior to doing it. This ensures a better level of trust and excellent communication exists during the consultation.

Testing at the Bedside
When examining the patient and discussing why they have presented to you, a full history should always be taken (Table 2.3). Further

Table 2.3 Systems That Should Be Discussed With a Patient When Undertaking a History Taking Exercise

History Taking	Main points to elicit
Personal Details	<ul><li>Name</li><li>Date of birth</li><li>Gender</li><li>Occupation</li><li>Source of the history</li></ul>
Presenting complaint	This should be in the patient's own words
History of presenting complaint	<ul><li>What the symptoms/signs are</li><li>The region of the body that is affected</li><li>When did the signs/symptoms commence?</li><li>Duration of signs/symptoms</li><li>Was the onset gradual or sudden?</li><li>Is there a history of trauma?</li><li>Have there been any exacerbating or relieving factors?</li><li>Any other associate features</li></ul>
Past Medical History	<ul><li>Medical e.g. hypertension, diabetes mellitus, asthma, hay fever, previous myocardial infarction, angina, epilepsy etc</li><li>Surgical, including any complications and indications for the surgery</li></ul>
Obstetric/gynecological/sexual health	<ul><li>Full obstetric history</li><li>Menstraul history</li><li>Birth controls</li><li>Sexual health record</li></ul>
Psychiatric	<ul><li>Type of illness</li><li>Duration</li><li>Hospitalisation</li><li>Diagnosis</li><li>Treatments</li></ul>
Family History	<ul><li>May help to draw a family tree</li><li>detail the age of the individual(s) and general health, and if there is a notable family history, how it has affected the relatives of the patient.</li><li>hypercholesterolaemia, hypertension, diabetes mellitus, asthma, hay fever, previous myocardial infarction, angina, epilepsy, genetic conditions etc.</li></ul>
Social History	<ul><li>Alcohol consumed</li><li>Smoking</li><li>Use of illicit drugs</li><li>Occupation</li><li>Relationship status</li><li>Exercise and dietary factors</li></ul>
Systems Review	<ul><li>Cardiovascular</li><li>Respiratory</li><li>Gastrointestinal</li><li>Head, eyes, ears, nose and throat</li><li>Neurological</li><li>Genitourinary</li><li>Musculoskeletal</li></ul>

The main points to discuss with them are highlighted.

details of the main features in a history taking can be found in the companion text to this resource in *Essential Clinically Applied Anatomy of the Peripheral Nervous system in the Limbs* (Rea, 2015).

Examination of the patient should be in relation to the signs and symptoms presenting, and any related features of any other systems which may be involved. Again, further details about a general clinical examination can be found in the companion text to this resource ("Essential Clinically Applied Anatomy of the Peripheral Nervous system in the Limbs" (Rea, 2015)). However, from here, the main features in an examination will only be given for the nerve which is being discussed in each subchapter and any cross-references which may be appropriate.

Testing of the olfactory nerve is often missed in routine clinical examination. As it is involved in the special sensation of smell, testing of this nerve is undertaken by using a substance with a recognizable smell.

Tip!

It is useful to evaluate the patency of the nasal passages bilaterally first. Asking the patient to breathe in through their nose while the examiner occludes one nostril at a time can do this. Always tell the patient what you are going to do before doing it!

- Tell the patient that you want to test their sense of smell and ask permission to do so. Use vanilla essence, coffee, orange peel, or lemon juice.
- Then do the following:
  1. Ask them to cover one nostril at a time and then close their eyes and present the testing substance to each nostril. The testing substance must not be visible to the patient.
  2. Ask the patient to report if they smell anything. This allows identification of the ability to detect an odor. Asking them to identify what is involves an "olfactory memory," that is, higher cortical functioning, only if they recognize the substance. *Do not* touch the patient when doing so, and place it within 15−30 cm from the nasal cavity.
  3. Do the same for the opposite nostril.

Tip!

Do not use any odors that are irritating, for example, menthol or ammonia, as these substances can stimulate the trigeminal nerve as well as the olfactory nerve resulting in a false-positive response.

Pathologies

A variety of pathologies can affect the olfactory nerve and can be thought of as falling into one (or perhaps several) category, as described in Table 2.4.

Table 2.4 Types of Pathologies That Can Affect the Olfactory Nerve Broken Down Into Broad Categories	
Local factors	• Allergic /atrophic/vasomotor rhinitis • Sinusitis • Asthma • Nasal polyps • Tumours • Exposure to toxic chemicals • Nasal trauma
Neurological	- Parkinson's disease - Alzheimer's disease - Multiple sclerosis - Epilepsy - Korsakoff's psychosis - Depression Schizophrenia
Neoplastic	**Intracranial** - Meningiomas - Frontal lobe glioma **Intranasal** - Adenocarcinoma - Squamous cell carcinoma - Papilloma
Endocrine	- Diabetes mellitus - Cushing's syndrome - Adrenocortical insufficiency - Pseudohypoparathyroidism
Nutritironal/other	- Vitamin B12 deficiency - Chronic renal failure - Herpes simplex Influenza

2.4.2 Optic Nerve

The second cranial nerve is the optic nerve and measures about 4 cm in length. It is an unusual cranial nerve as it actually develops from the diencephalon. That means that the structures which receive and transmit the visual information are extensions of the forebrain and are therefore CNS tracts formed by axons of the retinal ganglion cells. As such, and like the rest of the brain, the optic nerves are surrounded by the meninges (pia, arachnoid, and dura mater) and the cerebrospinal fluid (CSF)-filled subarachnoid space.

Visual input reaches the posterior aspect of the eye and the pathway begins with the photoreceptive cells called cones and rods in the retina, adjacent to the pigment epithelium. The cone cells are what we need for color vision and function best in rather bright light. There are approximately seven million cones in a human eye. The greatest density of cone cells is found within the *fovea centralis* of the *macula*. At this point, there is an absence of rod cells, and further away from the fovea centralis, there is a sharp decline in numbers of cones and an increase in rod cells.

Cones have three different pigments within them depending if they are sensitive to long, short, or medium wavelengths of light. Long wavelength cones, or L-cones, are sensitive to, and absorb, red light (564–580 nm). Short wavelength cones, or S-cones, are sensitive to, and absorb, blue light (420–440 nm). Medium wavelength cones, or M-cones, are sensitive to, and absorb, green light (534–545 nm). Therefore, in humans, we are known to have trichromatic vision. However, other vertebrates including birds, fish, and reptiles have four distinct cone photoreceptor cells (Okano et al., 1992; Bowmaker, 1998). On development of night activity by placental mammals, two visual pigments were lost, with primates then developing the third visual pigment, resulting in the current state of trichromatic vision (Jacobs, 1993).

Rod cells on the other hand are significantly greater in number (130 million) and function in low light intensity. Most of these cells are found on the outer aspect of the retina and are important for peripheral vision. They are structurally similar to the cone cells and are able to be incredibly efficient at absorbing light, and are therefore essential in night vision. Unlike cone cells, rod cells have only one light sensitive visual pigment so do not really play a role in color vision.

The responses of the photoreceptor cells (first-order neurons) are transmitted to the bipolar cell layer (second order neurons). These nerves have two processes—one coming from the rods and cones and the other projecting to the ganglion cell layer. From the ganglion cell layer, the impulses are then transmitted to the axons of the retinal ganglion cells. The axons of the retinal ganglion cells, which project centrally, are called the third-order neurons.

The central processes of the axons of the retinal ganglion cells join to form the optic nerve. The optic nerve passes backwards and in a medial direction. It passes through the posterior part of the orbital cavity, running through the optic canal.

Within the orbit, the optic nerve is closely related to the four recti muscles. The ciliary ganglion (from the oculomotor nerve) is found between the optic nerve and the lateral rectus.

As the optic nerve passes through the small *optic canal* (0.5 cm long), it contains the layers of the meninges, including the CSF-filled subarachnoid space. The central artery (from the ophthalmic artery) and vein of the retina travel through the meninges, passing to the anterior part of the optic nerve.

Within the cranial cavity, each optic nerve passes posterior and medially for approximately 1 cm to the *optic chiasma*. From this point, nerve fibers on the medial, or nasal, side of each retina cross the midline to the optic tract of the opposite side. However, the nerve fibers on the lateral, or temporal side, pass posteriorly but do not cross.

From each *optic tract*, visual information passes to the *lateral geniculate body* in the *midbrain*. From this point, the optic radiation carries visual input onwards to the optic radiation and then on to the occipital visual cortex. A small number of fibers for ocular and pupillary reflexes bypass the lateral geniculate body and pass straight to the pretectal nucleus and superior colliculus. Refer to Figures 2.1 and 2.2 for prosected images demonstrating relevant structures.

The fiber types present in the optic nerve are:

Special Sensory
• Visual information. It is a special visceral afferent nerve.

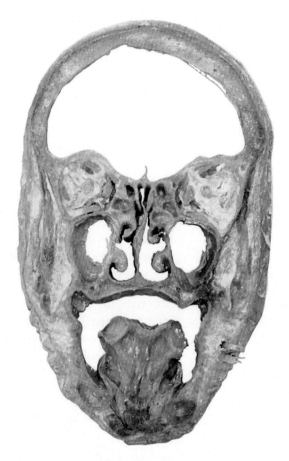

Figure 2.1 An unlabeled image to aid identification of the optic nerve.

Within the retina, there are 10 layers of cells involved in transmission of visual information from the (choroid) to the optic nerve:

1. Pigment cell layer
2. Layer of cones and rods
3. Outer limiting membrane
4. Outer nuclear layer
5. Outer plexiform layer
6. Inner nuclear layer
7. Inner plexiform layer
8. Ganglion cell layer
9. Nerve fiber layer
10. Inner limiting membrane

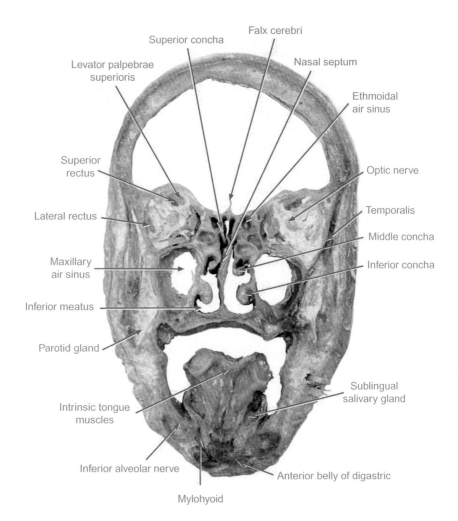

Figure 2.2 A fully labeled image of the position of the optic nerve in relation to other structures.

The branches of the optic nerve are as follows:

The retina and its projections can be divided up by a circle halved through the fovea based on its projections from either the nasal (medial) or temporal (lateral) halves. The lens inverts the visual images. Therefore in the left eye, the temporal half sees the left side of the world and the nasal half sees the right side of the world.

1. *Left eye—temporal side*
    a. Visual information passes to the right side of the left retina
    b. This passes to the left optic nerve

    c. It then crosses the optic chiasm

    d. The visual information then passes to the right lateral geniculate body

    e. Then onwards to the right visual cortices of the occipital lobe.

2. *Left eye—nasal side*

    a. Visual information passes to the left side of the left retina

    b. This passes to the left optic nerve

    c. *No* crossing over of this information occurs at the optic chiasm

    d. The visual information then passes to the left lateral geniculate body

    e. Then onwards to the left visual cortices of the occipital lobe.

3. *Right eye—temporal side*

    a. Visual information passes to the left side of the right retina

    b. This passes to the right optic nerve

    c. It then crosses the optic chiasm

    d. The visual information then passes to the left lateral geniculate body

    e. Then onwards to the left visual cortices of the occipital lobe

4. *Right eye—nasal side*

    a. Visual information passes to the right side of the right retina

    b. This passes to the right optic nerve

    c. *No* crossing over of this information occurs at the optic chiasm

    d. The visual information then passes to the right lateral geniculate body

    e. Then onwards to the right visual cortices of the occipital lobe.

Clinical Examination

As with all clinical examinations, introduce yourself to the patient, stating who you are and your purpose for meeting with the patient, especially if you are a student. Always take a comprehensive history before examining the patient. This will guide you to the most appropriate examinations that need to be conducted. Ensure you explain everything to the patient prior to doing it. This ensures a better level of trust and excellent communication exists during the consultation.

When clinically examining the optic nerve, five main features should be examined:

1. Visual acuity
2. Visual fields
3. Pupil size (reflexes)
4. Color assessment
5. Ophthalmoscopy

Visual Acuity

This is a measure of central vision and tests the ability of the patient to identify shapes and objects. Each eye *must* be tested separately. If a patient normally wears glasses or has contact lenses, perform the examination with these, and then without to ensure the patient's visual defect has been examined to identify any changes or worsening of their original condition.

A pocket visual acuity chart can be used for a general assessment, or a Snellen's Test Type chart can be used for more formal testing. This chart comprises different sized letters and is used to test *distant vision*. The Snellen chart is placed at 20 ft in the United States or 6 m away from the patient in all other countries. The point where a person with normal visual acuity can read to at 20 feet/6 meters is referred to as 20/20 vision (United States) or 6/6 vision (rest of the world). Therefore, a score for each eye has to be recorded when undertaking assessment of visual acuity.

Having 20/20 or 6/6 vision does not automatically mean perfect vision as it only tests central vision, and does not take into account peripheral visual fields, color awareness, or depth of perception.

In addition to this, near vision must also be examined. This is undertaken by using a card with sentences of different font sizes. Each paragraph has "points" where they are 1/72 inches apart. Therefore, N48 is the largest of the types and N5 is the smallest. This chart should be read at a comfortable reading distance of approximately 35 cm from each eye.

Visual Fields

This part of the examination of the optic nerve allows the examiner to assess both central and peripheral vision.

Confrontation Visual Field Testing (Donder's Test)

This involves standing or sitting (both patient and examiner) at the same eye level.

1. Ask the patient to cover one eye, for example, their right eye (examining the patient's left eye).
2. Ask the patient to remain looking at your eyes and to say "now" when the examiner's finger enters from out of sight into their peripheral vision.

3. Once the patient understands this, the examiner should cover their left eye with their left hand (opposite side to the patient).
4. Beginning with the examiners hand and arm fully extended, slowly bring your outstretched fingers centrally.
5. The examiner should bring their hand and outstretched fingers centrally.
6. The patient should then say "now" at the same time they see the examiners outstretched fingers and hand entering their field of vision.
7. This procedure should be undertaken once every 45° out of the 360° of peripheral vision.

Perimetry Test
This is a formal way of recording the visual fields. It systematically tests light sensitivity and identification by the patient on a predefined background.

Nowadays, this test is automated. The patient would sit down and look into a bowl-shaped instrument (a perimeter). The patient then has to stare at the center of the bowl and identify flashes of light. Each time the light flashes at varying points in the visual field (both central and peripheral), the patient presses a button to record when they see the flash of light.

A computer records when and where the light flashed and when the patient pressed the button indicating when they saw the light. A printed image is then obtained as to where the patient did not record seeing the visual flashing light. This then allows mapping of where the potential visual field defect lies.

Pupil Size
The pupils are approximately equal in size varying from 2−4 mm (bright light) to 4−8 mm (dark). When a light is shown into one eye, the pupil will constrict and this is referred to as the *direct response*. The pupil opposite to the side the light is shown into also constricts and this is referred to as the *indirect response*.

One acronym that is popular in clinical notes when documenting the pupil is *PERRLA*. This abbreviation stands for:

*P*upils
*E*qual

*R*ound and
*R*eactive to
*L*ight and
*A*ccommodation

Testing at the Bedside
1. Ask the patient to focus on something in the distance, perhaps on the other side of the room
2. Record the approximate diameter and shape of the pupil
3. SLOWLY move the light to the patient's eye from below the level of the eye, checking the pupillary response on the side the light is shown into, as well as the opposite one
4. Grade each pupils reaction to light from +1 to +4, with +4 being the fastest reaction
5. Then, ask the patient to focus from far to a near object, that is, if they were focusing on something at the opposite side of the room, ask them then to focus on, for example, a pen tip approximately 20 cm away from them.
6. Again, grade the pupillary response to this. The normal response is for miosis (pupillary constriction) when changing focus from a far to near object. It should be recorded in the same way as for the reaction to light (ie, +1 to +4)

Color Assessment
Color assessment is not typically examined in routine clinical practice. Typically, however, there are two versions of color vision deficiency:

1. Red−green deficiency. As the name suggests, this type of inherited condition affects the ability to differentiate red−green colors.
2. Blue−yellow deficient. This rare form of color "blindness" results in the inability to differentiate blue and yellow, where yellow appears as a grey or pale purple.

Typically, an optometrist will carry out this assessment using the Ishihara test, using multiple plates of 24 or 38 plates of multi-colored dots.

Ophthalmoscopy
The retina is unique in that it is the only part of the central nervous system that can be directly visualized from the external environment. The eye is a direct extension of the brain, and as such, carries with it the layers of the meninges. Any pressure increases within the cranial

cavity are therefore transmitted directly along the optic nerve and can be visualized with an ophthalmoscope.

Tip!

As well as a darkened room, it may help to instill mydriatic drops to dilate the pupil. Please check your local protocols and procedures, but this may include 1−2 drops of 0.5% tropicamide. This should be used approximately 15−20 min prior to examining the eye.

The examination should consist of the following:

1. Explain to the patient everything that you want to do. This will aid compliance.
2. Examine the lens and vitreous approximately 1 m away. Use high positive numbers on the ophthalmoscope.
3. Check for the red reflex which is the red glow from the choroid.
4. Cataracts will be seen as a black pattern obstructing this reflex.
5. Blood or loose floaters in the vitreous will be identified as black floaters.
6. Identify the position of the opacity in the eye.
7. When the retina is in focus, the optic disc should be examined.

Pathologies

Table 2.5 highlights the types of pathologies which can affect the optic nerve.

Table 2.5 Types of Common Pathologies Which Affect the Optic Nerve, and the Results This Has on the Eye	
Refractive errors	Myopia – short sightedness This is where the eyeball itself is too long in an antero-posterior direction a concave glass or contact lens is required. Hypermetropia – long sighted This is where the eye is too short in the antero-posterior direction A convex spectacle lens is required
Visual field defects	Lesion anterior to optic chiasm - Loss of vision on same side as damage Lesion at optic chiasm - Bitemporal hemianopia Lesion posterior to the optic chiasm - Loss of vision in visual field opposite to damage
Red eye	Acute glaucoma Corneal ulceration Acute iritis
Optic neuritis	Multiple sclerosis
Papilloedema	Many causes but requires urgent investigation to find the cause and treat the condition

2.4.3 Oculomotor Nerve

The oculomotor nerve arises from the anterior surface of the midbrain. Its nucleus is ventral to the aqueduct. From here, it passes through the dura mater to pass between the superior cerebellar and posterior cerebral arteries. From here, it passes to the middle cranial fossa in the lateral wall in the cavernous sinus. At this point, it also receives a few filaments from the cavernous plexus of the sympathetic nervous system, as well as a small communicating branch from the trigeminal nerve. Just as it enters the superior orbital fissure to gain access to the orbit, it divides into a superior and inferior division. The superior division innervates the levator palpebrae superioris and the superior rectus. The inferior division innervates the inferior oblique and the medial and inferior recti. The oculomotor nerve also carries fibers in it for parasympathetic innervation (or visceral efferent) via the ciliary ganglion to the *sphincter* (also called the pupillary constrictor or sphincter pupillae) and ciliary muscles. Refer to Figures 2.3 and 2.4 to demonstrate the nerve and related structures.

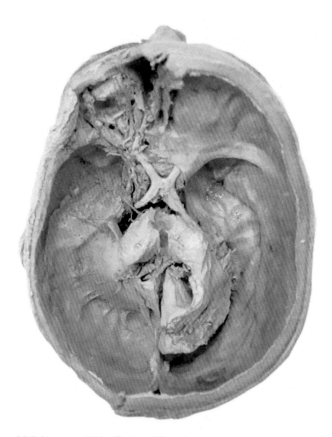

Figure 2.3 An unlabeled image to aid identification of the optic nerve.

The fiber types present in the oculomotor nerve are:

There are two types of fiber which are present in the oculomotor nerve—somatic motor and visceral motor. It is the somatic motor component which innervates many of the extraocular muscles of the eyeball. The visceral motor component of the oculomotor nerve innervates the parasympathetic part of this nerve namely the sphincter and ciliary muscles.

Somatic Motor

The cell bodies of the somatic motor part of the oculomotor nerve are found in the midbrain and they supply motor innervation to the medial, inferior and superior recti, levator palpebrae superioris, and the

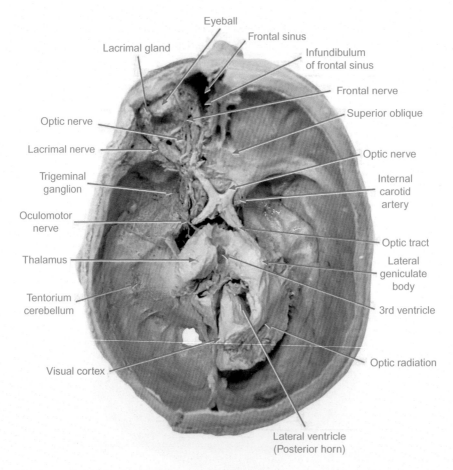

Figure 2.4 A fully labeled image of the position of the oculomotor nerve in relation to other structures.

inferior oblique. Table 2.6 summarizes the attachments and functions of each of these muscles in relation to movement of the eyeball/eyelid.

Table 2.6 The Extraocular Eye Muscles and the Movements Associated When They Contract		
Muscle	Attachments	Actions
Medial rectus	Annular tendon -> eyeball	Adduction of eyeball
Inferior rectus	Annular tendon -> eyeball	Depression and adduction of eyeball
Superior rectus	Annular tendon -> eyeball	Elevation and medial rotation of eyeball
Levator palpebrae superioris	Sphenoid bone -> upper tarsal plate	Elevation of eyeball
Inferior oblique	Maxilla -> eyeball	Abduction, elevation and lateral rotation

Visceral Motor

The visceral motor component of the oculomotor nerve supplies parasympathetic innervation (or visceral efferent) via the ciliary ganglion to the *sphincter* (also called the pupillary constrictor or sphincter pupillae) and ciliary muscles. Therefore, the presynaptic part is found in the midbrain with the postsynaptic fibers originating in the ciliary ganglion. The effect of stimulation of the visceral motor part of the oculomotor nerve results in constriction of the pupil and accommodation of the lens.

The branches of the oculomotor nerve are as follows:

There is a nice and simple way to remember what nerve innervates what muscle. If you use the following mnemonic to aid recollection of what muscle is innervated by which nerve, this will help in all settings when you need to recall the innervation of the extraocular muscles.

LR_6SO_4

This formulaic item mentioned above looks like something we encountered as some sulfur compound at school. Please do not confuse it as such, as it is a mnemonic to act as an aide memoire to the innervation of the extraocular muscles of the eye.

Simply, it means the following:

LR_6—*Lateral rectus*: innervated by the sixth cranial nerve (Abducens nerve)
SO_4—*Superior oblique*: innervated by the fourth cranial nerve (Trochlear nerve).

All other extraocular muscles are supplied by the oculomotor nerve, the third of the cranial nerves.

Clinical Examination

Two aspects of testing should be kept in mind when examining the integrity of the oculomotor nerve. The first is that the oculomotor nerve innervates muscles for eye movements, but also it innervates a component of the eye for the pupillary reflex.

As with all clinical examinations, introduce yourself to the patient, stating who you are and your purpose for meeting with the patient, especially if you are a student. Always take a comprehensive history before examining the patient. This will guide you to the most appropriate examinations that need to be conducted. Ensure you explain everything to the patient prior to doing it. This ensures a better level of trust and excellent communication exists during the consultation.

1. *Extraocular eye muscle movement*

   This simple test can be used to examine the integrity not only of the oculomotor nerve, but also the trochlear and abducens nerves.

   The following simple protocol can be followed when examining the extraocular eye muscles.

   a. Ask the patient to keep their head still during the examination, and only move their eyes.
   b. Ask the patient to follow, with their eyes only, the tip of a pen or finger at eye level.
   c. You should then move the pen (or fingertip) slowly in the horizontal plane from extreme left to extreme right.
   d. Go at a slow pace when doing this.
   e. When at the extreme left or right, with the examining object, stop!
   f. Observe for nystagmus.
   g. Then an H-shaped pattern with the examining object (pen or tip of finger) can be drawn in space, which the patient should follow *with their eyes only*
   h. All movements should be done *slowly* to assess for eye movements.

2. *Pupillary reflex*
   This can be tested in two ways:
   a. Ask the patient to follow your finger as you move it closer to their eyes. This will induce accommodation and, therefore, pupillary constriction.
   b. The swinging flashlight test can be used. This involves shining a pen torch into each pupil and observing both pupils for pupillary constriction. The purpose of this test is to identify if a relative afferent pupillary defect is present.

Advanced Testing
Abnormality in these initial tests or continuing clinical suspicion of oculomotor nerve damage should result in further investigation to identify any pathology which may be present. It may be necessary at this point to consult specialists for advice.

Further testing can be undertaken by examining brainstem nuclei which provide input to the oculomotor nerve. This would involve examining the patient's five aspects of ocular function namely fixation, saccadic movements, pursuit movements, compensatory movements, and opticokinetic nystagmus (Walker et al., 1990).

In terms of fixation, saccadic movements, and pursuit movements, observing the eye position and its related movements can assess these elements. For the compensatory movements, nystagmus should be assessed, as well as the doll's head maneuver.

Doll's Head Maneuver (Caloric Testing/Vestibular Caloric Stimulation)
The doll's head maneuver is also referred to as caloric testing or vestibular caloric stimulation. It is a test of the vestibule−ocular reflex and can be used to diagnose pathology in the peripheral vestibular system via eye movements. It is also a test used in brainstem death testing. Therefore, it tests numerous structures namely the labyrinthine structures, pontine lateral gaze centers, medial longitudinal fasciculus (crossed fibers tract which carries information about which direction the eyes should move connecting

the oculomotor, trochlear, and abducens nerves), and nuclei and peripheral nerves of the oculomotor and abducens nerves. It should be tested as follows:

1. *Always* inform the patient what you are going to do. Take time to explain this test, as it can be slightly unpleasant.
2. Warm water ($>44°C$) or cold water can be used. It may be easier in the clinical environment obtaining colder water.
3. *Always* check the integrity of the tympanic membrane first to ensure there is no perforation present.
4. Elevate the patient's head to approximately 30°. This ensures that only the horizontal semicircular canals will be stimulated.
5. Inject the cold (or warm water) into the auditory tube. In the conscious patient <1 ml may stimulate the response, while it may take up to 50 ml in the unconscious patient. This should only be done for <1 min.
6. The eyes should be observed over up to 3 min.

Response—This depends if the patient is conscious or not.

Conscious patient:

It depends if hot or cold water is used as the initial stimulus.

There are two components of nystagmus—fast and slow.

Cold Water
   The slow component of nystagmus is *towards* the side of injected cold water.
   The fast component of nystagmus is then *away* from the side of injected cold water.

Warm Water
   The slow component of nystagmus is *away* from the side of injected hot water.
   The fast component of nystagmus is *towards* the side of injected hot water

Tip!

There is a mnemonic which will help you to remember which direction the eyes will move in the fast component of nystagmus when undertaking the caloric test. It is by the following:

COWS
C—Cold
O—Opposite
W—Warm
S—Same

Unconscious patient:

Provided that the patient has a normal vestibular system and no brainstem pathology, no fast nystagmus is present, that is, if cold water is injected; there is only movement of the eye to the ear which has been injected with cold water.

Pathologies

The oculomotor nerve innervates the following muscles, and therefore results in the following eye movements (Table 2.7):

If there is an isolated oculomotor nerve palsy, but with intact trochlear and abducent nerves, the following will be seen of the eye:

1. The eye will appear to look "down and out", that is, due to the intact functions of the trochlear (lateral rotation and depression of eye) and abducens nerves (abducts the eye).
2. There will be pupillary dilation and it will be unresponsive (paralysis of sphincter muscle).
3. Ptosis will be present which means that there will be a droopy eyelid due to paralysis of levator palpebrae superioris.

Table 2.7 Summary of the Extraocular Eye Muscles	
Muscle	Movement
Inferior oblique	Elevation when adducted
Inferior rectus	Depression when abducted
Superior rectus	Elevation when abducted
Medial rectus	Adduction
Levator palpebrae superioris	Eyelid elevation
Sphincter muscle	Pupillary constriction

For diagnosing oculomotor nerve lesions, Warwick's scheme (1953) for the nerve complex can be used. From the clinical signs, a potential point of nerve damage of the oculomotor nerve could be identified before more in-depth investigation, for example, by CT/MRI scanning. This resulted in classifications based on if there was not a condition not representing an oculomotor nerve nuclear complex pathology, conditions that could potentially be oculomotor nerve nuclear complex in origin, and oculomotor nerve nuclear complex lesions.

Table 2.8 summarizes the possible pathologies which may be present to cause an oculomotor nerve dysfunction.

2.4.4 Trochlear Nerve

The fourth cranial nerve is the trochlear nerve. The trochlear nerve is a rather unusual cranial nerve in comparison to the others because:

- It is the smallest of the cranial nerves
- Longest intracranial course of the cranial nerves
- Only one of the cranial nerves to arise from the dorsal aspect of the midbrain
- One of only two nerves to decussate (the other is the optic nerve)

It supplies a single muscle—the superior oblique.

As a long slender structure, the trochlear nerve arises from the posterior (dorsal) aspect of the midbrain. It lies immediately caudal to the nucleus of the oculomotor nerve. The fibers of the trochlear nerve pass around the periaqueductal gray matter decussating at the level of the

Table 2.8 The Classifications of Pathologies Which May Affect the Oculomotor Nerve	
Vascular	- Aneurysm (of the posterior communicating, posterior cerebral or superior cerebellar artery) - Atherosclerotic changes +/−co-existing diabetes mellitus - Cavernous sinus thrombosis
Neurological	- Congenital oculomotor nerve paralysis - Space occupying lesion (intracranially) - Diabetic neuropathy
Trauma	- Base of skull fractures - Iatrogenic e.g. previous neurosurgery around the oculomotor nerve
Inflammation/infection	Linked to cavernous sinus thrombosis
Autoimmune	Possibly myasthenia gravi

superior medullary velum. This is a small area of white matter passing between the two superior cerebellar peduncles. It forms part of the roof of the fourth ventricle.

The trochlear nerve exits the midbrain just below the level of the inferior colliculus. It passes anteriorly through the middle cranial fossa in the outer aspect of the wall of the cavernous sinus. During its course to the superior orbital fissure, it is closely related to the superior cerebellar and posterior cerebral arteries.

As it enters the superior orbital fissure, it is closely related to the oculomotor and abducens nerves, and the ophthalmic and maxillary divisions of the trigeminal nerve. It then terminates in the muscle it supplies—the superior oblique. Refer to Figures 2.5 and 2.6 which demonstrates the nerve and related structures.

The fiber types present in the trochlear nerve are:

Somatic Motor

The trochlear nerve has somatic motor fibers which supplies a single muscle—the superior oblique. The tendon of the superior oblique passes through a pulley-like structure called the trochlea, hence the name trochlear nerve.

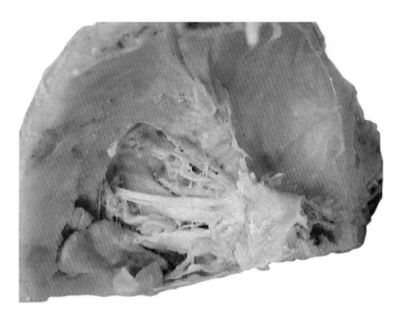

Figure 2.5 An unlabeled diagram to demonstrate the position of the trochlear nerve in relation to other structures.

The branches of the trochlear nerve are as follows:

Although there are no known branches of the trochlear nerve, it does receive connections with the medial longitudinal fasciculus (a nerve pathway for transmitting information about eye movements), tectobulbar tract (pathway transmitting information coordinating head and eye movements), and the corticonuclear (corticobulbar) tracts (transmitting information related to the motor functioning of non oculomotor cranial nerves).

Clinical Examination

As with all clinical examinations, introduce yourself to the patient, stating who you are and your purpose for meeting with the patient, especially if you are a student. Always take a comprehensive history before examining the patient. This will guide you to the most appropriate examinations that need to be conducted. Ensure you explain

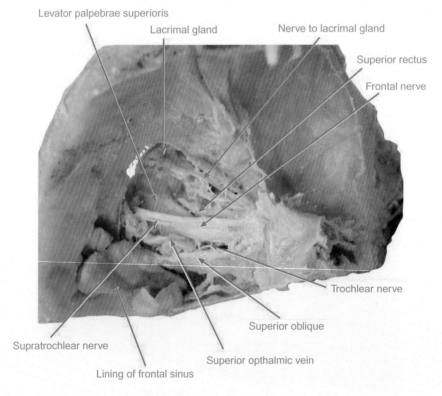

Figure 2.6 A labeled diagram to demonstrate the position of the trochlear nerve in relation to other structures.

everything to the patient prior to doing it. This ensures a better level of trust and excellent communication exists during the consultation.

The first thing to remember about the superior oblique is that it has a rotational movement both in the vertical plane and also the medial/lateral planes. The superior oblique muscle initially lies behind the eye; however, its tendon passes anteriorly before reaching the superior point of the eye. Therefore, the tendon of the superior oblique has two forces—one depressing the eye and one causing rotation of the eyeball to the nose.

However, the strength of each of these components depends on the way the eye is looking. When the eye is medially rotated, depression of the eyeball is increased. When the eyeball is laterally positioned, the rotation of it towards the nose increases. When the eyeball is in the neutral position, depression and rotation towards the nose work in approximately equal amounts.

Therefore, injury to the trochlear nerve will result in weakness of the eyeball to move downwards with diplopia (double vision). The patient will notice this vertical diplopia most when looking downwards, for example, when walking down stairs or reading a book.

Testing at the Bedside
Testing of the trochlear nerve is undertaken when assessing the oculomotor nerve and the abducens nerves at the same time. It involves testing of all of the extraocular muscles. Therefore, the procedure for testing the trochlear nerve is as follows:

1. Ask the patient to keep their head still during the examination
2. With their eyes only, they should follow the tip of your finger (or pen torch, pencil, etc.)
3. The examiner should then move the object in the horizontal plane from extreme left to extreme right
4. This should be done *slowly*
5. When at the extreme left or right, with the examining object, stop!
6. Observe for nystagmus
7. Then an H-shaped pattern with the examining object can be drawn in space, which the patient should follow *with their eyes only*
8. All movements should be done *slowly* to assess for eye movements.

Table 2.9 Types of Pathologies Which Can Affect the Trochlear Nerve

Infection	Sinusitis e.g. of the sphenoid
Iatrogenic	Neurosurgical procedures
Neurological	Progressive supranuclear palsy Multiple sclerosis or other demyelinating disease Myasthenia gravis
Trauma	Trauma to the head e.g. from a road traffic accident
Vascular	Hypertension Aneurysms Microvascular disease e.g. because of diabetes mellitus Haemorrhage intracranially Arteriovenous malformation
Neoplastic	Space occupying lesion within the cranial cavity
Congenital	Isolated pathology of the trochlear nerve

Pathologies

The long intracranial course of the trochlear nerve makes it particularly vulnerable to injury from blunt head trauma, or any pathologies which cause an increase in intracranial pressure. Brainstem lesions that damage the trochlear nerve nucleus are not easily identifiable clinically. Table 2.9 summarizes the main causes of a trochlear nerve palsy.

2.4.5 Trigeminal Nerve

The trigeminal nerve arises from the lateral aspect of the pons comprising a large sensory root and a smaller motor root. Cell bodies of the trigeminal nerve are located in the trigeminal ganglion with a lesser amount in the mesencephalic trigeminal nucleus.

It is the peripheral processes of the ganglion that forms the ophthalmic and maxillary nerves and the sensory part of the mandibular nerve. In addition, within the mandibular nerve, proprioceptive fibers are present from the mesencephalic nucleus.

Central processes of the trigeminal ganglion enter the pons and then pass to the spinal and pontine trigeminal nuclei. It is the large fibers for discriminative touch that terminate in the pontine trigeminal nucleus. Indeed, the pontine trigeminal nucleus is referred to as the chief or principal sensory nucleus.

However, a smaller number of fibers pass caudally towards the spinal cord to the spinal trigeminal nucleus. The spinal trigeminal nucleus is responsible for conveying information related to light touch, pain, and temperature.

The information conveyed in the spinal trigeminal tract also includes input from the outer aspect of the ear, posterior one-third of the tongue (mucosa of), pharynx, and larynx. This means there is an input related to sensation also from these sites and is related to the facial, glossopharyngeal, and vagus nerves, respectively. In the sensory root and the spinal cord portion of the trigeminal nerve, there is a spatial arrangement of the fibers. The mandibular fibers are initially ventral and the ophthalmic fibers dorsal, with the maxillary fibers lying between these. On approach to the brainstem, there is a rotation of the fibers to lie the opposite way, that is, the mandibular fibers end up dorsal and the ophthalmic fibers ventral, and again the maxillary fibers sandwiched between the two.

In addition to this, the mesencephalic trigeminal nucleus extends from the pontine trigeminal nucleus to the midbrain. This has two processes—a central part and a peripheral portion. The peripheral branches of the mesencephalic trigeminal nucleus pass within the mandibular nerve and terminate in proprioceptive receptors beside the teeth of the mandible and in the muscles of mastication (ie, temporalis, masseter and the pterygoid muscles). Occasional fibers also pass to the maxillary division ending in the hard palate adjacent to the teeth of the maxilla.

The central branches of the mesencephalic trigeminal nucleus terminate in either the motor nucleus of the trigeminal nerve or the reticular formation (and then onwards to the thalamus).

The bulk of the motor root of the trigeminal nerve contains fibers from the trigeminal motor nucleus. This nucleus, found medial to the chief or principal sensory nucleus (ie, pontine trigeminal nucleus), supplies the muscles of mastication (temporalis, masseter, and the pterygoids (lateral and medial)). It also supplies the anterior belly of digastric, mylohyoid, tensor veli tympani, and tensor tympani. The trigeminal motor nucleus receives afferent information from the corticobulbar tract (that white matter pathway connecting the cerebral cortex to the brainstem). Afferent information also arrives from the sensory trigeminal nuclei. This pathway deals with the stretch reflex and the jaw-opening reflex.

On leaving the brainstem, the motor fibers of the trigeminal nerve pass below the ganglion along the floor of Meckel's cave (named after the German anatomist Johann Friedrich Meckel, the Elder (to avoid confusion with his famous grandson, also anatomists)). These fibers are only present in the mandibular division of the trigeminal nerve and become related to the sensory fibers as the whole nerve passes through the foramen ovale. It goes on to supply the muscles of mastication, and some of the smaller muscles previously described. Refer to Figures 2.7 and 2.8 to demonstrate the nerve and related structures.

OPHTHALMIC NERVE

The ophthalmic nerve is the first division of the trigeminal nerve. The ophthalmic division of the trigeminal nerve, also referred to as the ophthalmic nerve, is a purely sensory (afferent) nerve. It is the smallest division of the trigeminal nerve. It runs forward in the lateral wall of the cavernous sinus below the oculomotor and trochlear nerves. It divides into the frontal, lacrimal, and nasociliary nerves, which enter the orbital cavity through the superior orbital fissure. It does not supply pharyngeal arch origin structures as it is derived from the paraxial mesoderm. In general, it supplies the skin and mucous membranes of the head and nose at the front. It supplies the skin of the face above the level of the orbit, but extending down to the tip of the nose only

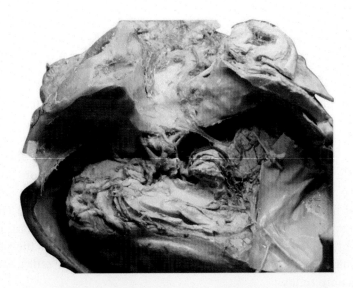

Figure 2.7 An unlabeled image to demonstrate the position of the trigeminal nerve in relation to other structures.

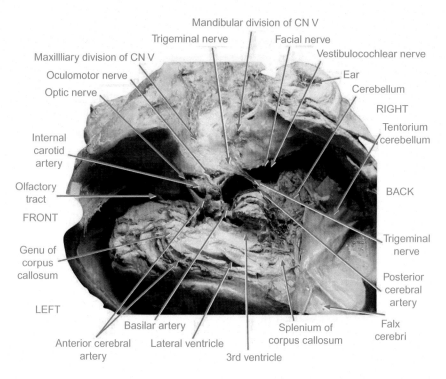

Maxillliary division of CN V
Oculomotor nerve
Optic nerve
Internal carotid artery
Olfactory tract
FRONT
Genu of corpus callosum
LEFT
Trigeminal nerve
Mandibular division of CN V
Facial nerve
Vestibulocochlear nerve
Ear
Cerebellum
RIGHT
Tentorium cerebellum
BACK
Trigeminal nerve
Posterior cerebral artery
Falx cerebri
Anterior cerebral artery
Basilar artery
Lateral ventricle
3rd ventricle
Splenium of corpus callosum

Figure 2.8 A labeled image to demonstrate the position of the trigeminal nerve in relation to other structures.

on the anterior aspect. It does not supply the lateral aspect of the nose. These features are important in clinical testing and pathology involving a single branch of the trigeminal nerve, which will be discussed later. As well as supplying the skin of the forehead and the upper eyelid, it also supplies the cornea, paranasal sinuses, and nasal mucosa. The specific branches of the ophthalmic nerve are as follows:

a. *Frontal nerve*—this terminates as the supraorbital and supratro-chlear nerves innervating the skin of the scalp over the area of fron-talis and the frontal bone, frontal sinuses, and the upper eyelid.
b. *Lacrimal nerve*—provides sensory innervation to the lacrimal gland, conjunctiva, and the upper eyelid.
c. *Nasociliary nerve*—supplies the nasal mucous membrane, paranasal sinuses, and is also involved in afferent loop of the corneal blink reflex.
d. *Meningeal branch (tentorial nerve)*—arising from the intracranial part of the ophthalmic nerve. This branch supplies the falx cerebri (supratentorial portion) and the tentorium cerebelli.

Lacrimal Nerve

The lacrimal nerve enters the orbit through the superior orbital fissure, above the muscles of the eyeball. It proceeds along the upper border of the lateral rectus and ends at the front of the orbit by giving branches to the lacrimal gland, the skin of the upper eyelids, and the conjunctiva. The lacrimal nerve communicates in the orbit with the zygomatic nerve, and by this route, some secretory fibers are then brought to the lacrimal gland.

Frontal Nerve

The frontal nerve enters the orbit through the superior orbital fissure, above the muscles of the eyeball, and passes anterior on the levator palpebrae superioris. The frontal nerve divides into two branches, although the point of division is highly variable.

a. Supraorbital nerve
   The supraorbital nerve is the direct continuation of the frontal nerve. It exits the orbit through the supraorbital notch or foramen and is distributed to the forehead and the scalp, upper eyelid, and also the frontal sinus.
b. Supratrochlear nerve
   The supratrochlear nerve is considerably smaller than the supraorbital nerve. It exits the orbit at the medial end of the supraorbital margin. It innervates the forehead and also the upper eyelid.

Nasociliary Nerve

The nasociliary nerve is the sensory nerve to the eye. It enters the orbit through the superior orbital fissure, inside the cone formed by the muscles of the globe. It is on a lower plane, therefore, than the lacrimal and frontal nerves. It lies between the two divisions of the oculomotor nerve. It passes anteriorly inferior to the superior rectus and crosses the optic nerve with the ophthalmic artery. At the medial aspect of the orbit, it lies between the superior oblique and the medial rectus. It is continued as the anterior ethmoidal nerve.

The nasociliary nerve gives off several smaller branches:

1. Communicating branch
   This communicating branch passes to the ciliary ganglion (see Abducens Nerve).

2. Long ciliary nerves
   These can exist either singly or two nerves and convey sympathetic fibers to the dilator pupillae and afferent fibers from the uvea and cornea.
3. Infratrochlear nerve
   The infratrochlear nerve passes to the eyelids, skin of the nose, and the lacrimal sac.
4. Posterior ethmoidal nerve
   The posterior ethmoidal nerve is frequently absent. When present it provides sensory innervation to the posterior ethmoidal and sphenoidal sinuses.
5. Anterior ethmoidal nerve
   The anterior ethmoidal nerve is seen as the continuation of the nasociliary nerve. The anterior ethmoidal nerve passes through the anterior ethmoidal foramen entering into the anterior cranial fossa. It passes into the nasal cavity dividing into internal nasal branches. These branches supply the walls of the nasal cavity. One of the branches passes to the skin of the nose as an external nasal branch.

In its course, the nasociliary nerve, together with its continuation, the anterior ethmoidal nerve, traverses in succession the middle cranial fossa, the orbit, the anterior cranial fossa, nasal cavity, and the external aspect of the nose.

MAXILLARY NERVE
The maxillary division of the trigeminal nerve, also referred to as the maxillary nerve, is a purely sensory (afferent) nerve. It is the medium-sized branch of the trigeminal nerve between the smaller ophthalmic division and the largest mandibular division. After emerging from the trigeminal ganglion, it passes to the pterygopalatine fossa, passing to the posterior surface of the maxilla before passing through the foramen rotundum and entering the orbit through the inferior orbital fissure, running here to terminate on the anterior aspect of the skull at the infraorbital foramen. The maxillary nerve also passes through the cavernous sinus. In general, it supplies the teeth of the maxilla, skin from the lower eyelid above to the superior aspect of the mouth below, as well as the nasal cavity and the

paranasal sinuses. The specific branches of the maxillary nerve are as follows:

Middle Meningeal Branches

The middle meningeal branches or the meningeal nerve arises from the maxillary nerve close to the foramen rotundum. It also receives a branch from the internal carotid sympathetic plexus and then joins with the frontal branch of the middle meningeal artery. The middle meningeal nerve innervates the dura mater within the middle cranial fossa. Some of its anterior fibers will go on to innervate the anterior cranial fossa.

Ganglionic Branches

The ganglionic branches of the maxillary nerve unite the trigeminal nerve with the pterygopalatine ganglion, found in the pterygopalatine fossa. The pterygopalatine fossa is a depression deep within the infratemporal fossa. Found behind the maxilla. It is found between the maxillary tuberosity and the pterygoid process, near to the orbit. The ganglionic branches of the maxillary nerve contain the secretomotor fibers destined for the lacrimal gland. Also contained within the ganglionic branches are sensory fibers from the periosteum of the orbit, as well as mucous membranes of the pharynx, palate, and nose.

Zygomatic Nerve

The zygomatic nerve starts within the pterygopalatine fossa and enters the orbit via the inferior orbital fissure dividing into two named branches, that is, zygomaticotemporal and zygomaticofacial nerves.

The zygomaticotemporal nerve runs along the lower outer aspect of the orbit and provides a branch to the lacrimal nerve. It then passes through a small canal in the zygomatic bone and then arrives into the temporal fossa. It passes superiorly between the bone and the temporalis muscle. It will then go through the temporal fascia just a couple of centimeters superior to the zygomatic arch and innervates the skin of the temple region. It anastomoses with the facial nerve and also with the auriculotemporal nerve forms the mandibular division of the trigeminal nerve. When it pierces the temporal fascia, it also sends a small branch between the two layers of temporal fascia to reach the outer aspect of the eye.

The zygomaticofacial nerve runs along the lower outer aspect of the orbit, and arrives onto the surface of the face through a foramen in the zygomatic bone. It then passes through the orbicularis oculi and inner-vates the skin over the prominence of the cheek. It ramifies with the zygomatic branch of the facial nerve and also the palpebral branches from the maxillary nerve of the trigeminal nerve.

Alveolar Nerves
The alveolar nerves are also referred to as the dental nerves and there are typically three sets superiorly which arise from the maxillary nerve—*anterior, middle,* and *posterior.* There is also an *inferior alveolar (dental) nerve,* but this arises from the mandibular division of the trigeminal nerve, and is a single nerve that terminates as the mental nerve upon exiting from the mental foramen of the mandible.

The superior alveolar (dental) nerve arises from the maxillary nerve prior to it leaving the pterygopalatine fossa, or in the infraorbital groove, or indeed the infraorbital canal. It gives off the anterior, middle, and posterior superior alveolar (dental) nerves).

The anterior superior alveolar (dental) nerve arises from the outer aspect of the infraorbital nerve at approximately mid-way along the infraorbital canal. It passes inferior to the infraorbital foramen, running medially towards the nose. It then passes inferiorly and splits into its branches which will innervate the canine and the incisor teeth. It provides innervation to the superior dental plexus, but also provides a small branch to the anterolateral wall of the nasal cavity.

The middle superior alveolar (dental) nerve originates form the infraorbital nerve as it runs along the infraorbital groove. It passes anteroinferiorly within the outer wall of the maxillary sinus. It also unites with the superior dental plexus and it gives off small branches to innervate the upper premolar teeth.

The posterior superior alveolar (dental) nerve comes from the max-illary nerve within the pterygopalatine fossa. It passes inferiorly and anteriorly to go through the maxilla's infratemporal aspect and then innervates the maxillary sinus. The posterior superior alveolar (dental) nerve innervates the molar teeth as well as the upper gums and the adjacent part of the cheek.

Palatine Nerves

The palatine nerves innervate the nasal cavity, the tonsil, roof of mouth, and the soft palate. There are three divisions of the palatine nerves—anterior, middle, and posterior.

The greater (anterior) palatine nerve passes through the greater palatine canal and enters the hard palate via the greater palatine foramen. It innervates the mucous membranes, gums, and the glands of the hard palate. It also communicates anteriorly with the nasopalatine nerve.

The lesser palatine nerves (middle and posterior) enter the greater palatine canal and go through the lesser palatine foramina innervating the tonsil, soft palate, and uvula.

Pharyngeal Branch

The pharyngeal (pterygopalatine) nerve originates from the posterior aspect of the pterygopalatine ganglion. Along with the pharyngeal branch from the maxillary artery, it innervates the mucous membranes that line the nasopharynx posterior to the auditory tube.

Infraorbital Nerve

Once the maxillary nerve passes into the infraorbital canal, it is then referred to as the infraorbital nerve. The infraorbital nerve innervates the mucous membrane and the skin of the middle aspect of the face. It has been shown that the cutaneous distribution of the infraorbital nerve is rather extensive. Averaging almost 20 branches, the infraorbital nerve had a cutaneous distribution of almost 26 cm^2 (Hwang et al., 2004).

Inferior Palpebral Nerve

The inferior palpebral or palpebral branches from the maxillary nerve passes deep to the orbicularis oculi. It then goes through this muscle to innervate the skin of the lower eyelid, and anastomoses with the facial and the zygomaticofacial nerves close to the outer aspect of the angle of the eye.

Superior Labial Nerve

The superior labial branches of the maxillary nerve are large and plentiful. They pass inferiorly behind the levator labii superioris. They innervate the skin of the front of the cheek, mucous membranes of the mouth, skin of the upper lip, and the labial glands. The facial nerve anastomoses with these branches thus forming the infraorbital plexus.

MANDIBULAR NERVE

The mandibular division of the trigeminal nerve, also referred to as the mandibular nerve, is a mixed sensory and branchial motor nerve. It is also the largest of the three branches of the trigeminal nerve. The sensory root arises from the lateral aspect of the ganglion, with the motor division lying deeper. In general, the mandibular nerve supplies the *lower face* for sensation over the *mandible*, including the *attached teeth*, the *TM joint*, and the *mucous membrane of the mouth* as well as the *anterior two-thirds of the tongue* (the posterior one-third is supplied by the *glossopharyngeal nerve*). It also supplies the *muscles of mastication* which are the *medial* and *lateral pterygoids, temporalis*, and *masseter*. It also supplies some smaller muscles namely the tensor veli tympani, tensor veli palatini, mylohyoid, and the anterior belly of digastric.

The mandibular nerve enters the infratemporal fossa and passes through the foramen ovale in the sphenoid bone, and divides at that point into a smaller anterior and a larger posterior trunk. The main trunk gives off two branches at this point. The first is a *meningeal branch* and passes through the foramen spinosum to receive innervation from the meninges of the middle cranial fossa. The second small branch, a muscular branch, which supplies the *medial pterygoid* and also a twig to the otic ganglion to supply the tensor veli palatini and the tensor tympani. Two main divisions arise from the main trunk of the mandibular nerve after these two smaller branches have been given off: an *anterior* and *posterior division*.

The otic ganglion is situated in the infratemporal fossa immediately below the foramen ovale, medial to the mandibular nerve (from the trigeminal nerve), lateral to the tensor veli palatine, in front of the middle meningeal artery, and behind the medial pterygoid muscle. The fibers connected with the ganglion are generally described as its roots. The parasympathetic (motor) root is the lesser petrosal nerve. These preganglionic fibers derived from the glossopharyngeal nerve synapse in the ganglion, and are the only fibers to do so. The postganglionic fibers pass to the auriculotemporal nerve. They are secretory to the parotid gland. A sympathetic root is derived from the plexus on the middle meningeal artery. These fibers are postganglionic (arising in the superior cervical ganglion). They merely pass through the otic ganglion and, by way of the auriculotemporal nerve, supply the blood vessels of the parotid gland. An efferent root comes from the nerve to the medial pterygoid muscle. These fibers pass through the ganglion and are said to supply the tensor tympani and the tensor veli palatine.

Occasionally, some taste fibers from the anterior two-thirds of the tongue can pass through the otic ganglion, which they reach by a communication from the chorda tympani and leave by a communication to the nerve of the pterygoid canal.

The branches of the mandibular nerve are now given below.

Meningeal Branch

The meningeal branch of the mandibular nerve passes into the skull from the mandibular nerve via the foramen spinosum accompanying the middle meningeal artery. It then divides into two branches—one running anteriorly and the other posteriorly. These branches innervate the *dura mater*, but also the anterior branch anastomoses with the meningeal branch form the maxillary nerve. The posterior branch of the meningeal branch from the mandibular nerve also provides innervation to the *mastoid air cells*.

Nerve to Medial Pterygoid

The nerve to medial pterygoid, or medial pterygoid nerve, gives two branches innervating the *tensor tympani* and the *tensor veli palatini*. It then passes into the deep aspect of the *medial pterygoid muscle*, thus innervating it.

Masseteric Nerve

The masseteric nerve passes posterior to the tendon of temporalis; this branch approaches the masseteric muscle on its deep aspect.

Deep Temporal Nerves

The deep temporal nerves exist as two branches that have an anterior and posterior division. Sometimes, a third (intermediate) branch may be found. The deep temporal nerves innervate the *temporalis*.

Lateral Pterygoid Nerve

The lateral pterygoid nerve branch enters the deep surface of the *lateral pterygoid muscle* to innervate it.

Buccal Branches

Tip!

DO NOT CONFUSE THIS WITH THE BUCCAL BRANCH OF THE FACIAL NERVE WHICH CONVEYS MOTOR INFORMATION TO BUCCINATOR.

This buccal branch from the anterior division of the mandibular nerve passes between the lateral pterygoid heads then inferior to the temporalis tendon passing to the buccal membrane to receive sensory information from that site, skin over the cheek, as well as the second and third molars. Indeed, some of the branches of the buccal nerve from the marginal mandibular nerve anastomose with the buccal branches of the facial nerve.

Auriculotemporal Nerve
The auriculotemporal nerve has two roots and is closely related to the middle meningeal artery. It passes posterosuperiorly behind the TMJ. It is closely related to the superficial temporal vessels and gives off secretomotor fibers to the parotid gland, before reaching the temporal region and receives sensory information from here, as well as the superior half of the pinna and the external auditory meatus.

Lingual Nerve
The lingual nerve carries two parts of information—that related to the trigeminal nerve for sensation from the tongue (general somatic afferent) but also carries with it a branch from the facial nerve—the *chorda tympani nerve* for special sensory fibers of taste from the *front two-thirds of the tongue*, but also *preganglionic parasympathetic fibers to the submandibular ganglion*. The lingual nerve first passes below the lateral pterygoid muscle, receives the chorda tympani nerve, and then passes between the medial pterygoid and then passes towards the tongue.

Inferior Alveolar Nerve
The inferior alveolar nerve is a dual motor and sensory nerve. It passes on the medial aspect of the lateral pterygoid, and just before entering the mandibular foramen, it gives rise to the motor branches to the *mylohyoid* and *anterior belly of digastric* muscles. After it enters the mandibular foramen, it supplies all the *lower teeth* and the *alveolar ridges*. As it passes anteriorly, it then exits the mandible via the mental foramen and becomes the *mental nerve*. This nerve supplies sensation to the *skin over the chin*. This nerve is crucial in dental practice, as it is the nerve that is anaesthetized as it enters the mandibular foramen to provide complete nerve block if procedures are to be undertaken on the lower teeth, or related structures.

Clinical Examination

As with all clinical examinations, introduce yourself to the patient, stating who you are and your purpose for meeting with the patient, especially if you are a student. Always take a comprehensive history before examining the patient. This will guide you to the most appropriate examinations that need to be conducted. Ensure you explain everything to the patient prior to doing it. This ensures a better level of trust and excellent communication exists during the consultation.

There are two aspects that can be tested for clinically with the trigeminal nerve—the motor function of the nerve and if sensation is intact.

1. Ensure you take a detailed clinical history first.
2. *Always* tell the patient what you will be doing and what you expect them to do in helping elicit any signs and/or symptoms.
3. Observe the skin over the area of temporalis and masseter first to identify if any atrophy or hypertrophy is obvious.
4. First, palpate the masseter muscles while you instruct the patient to bite down hard. Also note masseter wasting on observation. Do the same with the temporalis muscle.
5. Then, ask the patient to open their mouth against resistance applied by the instructor at the base of the patient's chin.
6. To assess the stretch reflex (jaw jerk reflex), ask the patient to have their mouth half open and half closed. Place an index finger onto the tip of the mandible at the mental protuberance, and tap your finger briskly with a tendon hammer. Normally this reflex is absent or very light. However, for patients with an upper motor neuron lesion, the stretch reflex (jaw jerk reflex) will be exaggerated.
7. Also ask the patient to move their jaw from side to side.
8. Next, test gross sensation of the trigeminal nerve. Tell the patient to close their eyes and say "sharp" or "dull" when they feel an object touch their face. Allowing them to see the needle, brush, or cotton wool ball before this examination may alleviate any fear. Using the needle, brush, or cotton wool, randomly touch the patient's face with the object. Touch the patient above each temple, next to the nose, and on each side of the chin, all bilaterally. You must test each of the territories of distribution of the ophthalmic, maxillary, and mandibular nerves.

9. Ask the patient to also compare the strength of the sensation of both sides. If the patient has difficulty distinguishing pinprick and light touch, then proceed to check temperature and vibration sensation using the vibration fork. You can heat it up or cool it down in warm or cold water, respectively.

10. Finally, test the corneal reflex (blink reflex). You can test it with a cotton wool ball rolled to a fine tip. Ask the patient to look at a distant object and then approaching laterally, touching the cornea (and not the sclera) looking for the eyes to blink. Repeat this on opposite eye. If there is possible facial nerve pathology on the side that you are examining, it is imperative to observe the opposite side for the corneal reflex.

Some clinicians omit the corneal reflex unless there is sensory loss on the face elicited from the history or examination, or if cranial nerve palsies are present at the pontine level. It is best to ensure a complete clinical examination is undertaken, however, especially if there is a possible pathology of the trigeminal nerve.

Clinical Applications

1. *Inferior alveolar nerve block*

   This is one of the most common dental procedures undertaken. It is where the inferior alveolar nerve is anaesthetized just before it enters the mandible at the mandibular foramen. When the anesthetic agent takes effect, it results in anesthesia of the lower teeth and related gingivae, but also the skin of the chin and the lower lip, as the inferior alveolar nerve terminates as the mental nerve as it exits the mental foramen.

   Following a nerve block of the inferior alveolar nerve for dental procedures of the lower teeth and surrounding structures, loss of sensation to the tongue can result. This is due to the fact that the lingual nerve is found very close to where the inferior alveolar nerve is anaesthetized as it enters the mandibular foramen. As the lingual nerve supplies sensation to the anterior two-thirds of the tongue, it can also be anesthetized resulting in a temporary loss of sensation to that portion of the tongue.

2. *Infraorbital nerve block*

   The infraorbital nerve may be blocked for a variety of reasons as shown in Table 2.10.

Table 2.10 Site of Anesthetic Injection and the Reasons for Doing This, Including the Anatomical Territory Which Will Be Anesthetized

Site of anesthesia	Reasons
Supraperiosteal infiltration	Anesthesia of maxillary dentition for a single tooth or soft tissue procedure
Periodontal ligament	Adjunct to the supraperiosteal injection Used as a supplement to ensure adequate anesthesia
Posterior superior alveolar nerve	Anesthetizing the maxillary molar teeth up to 1st molar
Middle superior alveolar nerve	Anesthetizing the maxillary premolar teeth or the mesiobuccal root of the 1st molar
Anterior superior alveolar nerve/ infraorbital nerve	Anesthetizing the maxillary central and lateral incisors and canine teeth
Greater palatine nerve	Anesthetizing the palatal aspect of the maxillary premolar and molar teeth
Nasopalatine nerve	Anesthetizing the lingual aspect of multiple anterior teeth

Pathologies

Herpes Zoster

The varicella zoster virus causes herpes zoster (shingles), the most common sensory abnormality of the face and scalp. It manifests in childhood as chicken pox (vesicular skin rash), but the virus is never eliminated from the body. Following the initial acute illness, it then remains dormant in the nerve cell bodies, including the trigeminal ganglion. Years or decades after the initial episode of chicken pox, the virus "wakens up" again and passes down the nerve axons to cause a viral infection of the skin again. This process can take a few days, and there is no obvious reason for it to flare up though most patients are older than 50 years who get herpes zoster. If a patient is younger than 50 years who has herpes zoster or affects more than one dermatome, an immunodeficiency should be suspected.

It results in a burning pain, itching (which can be severe), and a vesicular skin eruption. It affects one of the divisions of the trigeminal nerve, that is, ophthalmic, maxillary, or mandibular nerves. It can be a very serious condition if the ophthalmic division is affected as it can result in corneal ulceration, which can threaten the integrity of the eye, and therefore sight. Therefore, if ophthalmic herpes zoster is suspected, an ophthalmologist must be consulted immediately.

In terms of treating herpes zoster, topical antiviral treatment has been shown to be ineffective. Oral acyclovir has been shown to reduce the duration of the signs and symptoms, and reduce the rate and severity of complications related to ophthalmic herpes zoster.

The most common complication from herpes zoster is postherpetic neuralgia. Other complications of herpes zoster relate to ocular, and surrounding structures, therefore, input from an ophthalmologist is essential.

Trigeminal Neuralgia

Trigeminal neuralgia or tic douloureux (Fr. painful twitch) has a prevalence of 0.1−0.2 per 1000 and an incidence ranging from 4 to 5 per 100,000/year up to approximately 20 per 100,000/year after the age of 60 (Manzoni and Torelli, 2005). It affects twice as many women than men and rare below the age of 40 (NHS[2]). In this condition, it causes excruciating pain in one of the territories of distribution of the trigeminal nerve, with the maxillary nerve to be most commonly affected, followed by the mandibular nerve, then less common, the ophthalmic nerve. Generally, it affects only one side of the face, and the most common cause is pressure on the trigeminal nerve within the skull.

The condition presents as severe facial pain described by patients as "stabbing, shooting, excruciating, or burning" (Trigeminal Neuralgia Association United Kingdom[3]). Touching the affected area, even by light touch, often sets off the facial pain. It can last less than a second up to a couple of minutes, and there may be pain-free periods for months or years. It can be a very debilitating condition as day-to-day activities can bring on the pain, for example, the wind blowing on the face, chewing, speaking, eating and drinking, and shaving.

There are three different types of trigeminal neuralgia, types 1, 2 and symptomatic trigeminal neuralgia referred to as TN1, TN2, and STN respectively. TN1 is the classical form where the pain only occurs occasionally and is not constant. There is no identifiable cause for TN1. TN2 is referred to as atypical trigeminal neuralgia as the pain is more constant and is like a throbbing sensation. STN is where there is an identifiable cause, for example, multiple sclerosis.

[2]http://www.nhs.uk/conditions/trigeminal-neuralgia/Pages/Introduction.aspx (Accessed January 7, 2014).
[3]http://www.tna.org.uk (Accessed September 28, 2015).

In most patients the antiepileptic agent carbamazepine can be given. Although it is an antiepileptic agent, it works by reducing nerve impulses, hence dulls or eliminates the pain of the acute attack. If medical treatment does not work, it may be necessary to consider surgery for those patients severely affected by trigeminal neuralgia, yet unresponsive to medication.

Potential procedures, which can be undertaken if medication is ineffective, is thermocoagulation of the nerve endings, balloon compression, electric currents applied to the branch of the trigeminal nerve, peripheral radiofrequency, or glycerol injection of the trigeminal nerve branch.

Many cases of trigeminal neuralgia with no systemic cause are due to the presence of a small aberrant artery pressing against a branch of the trigeminal nerve. Microvascular decompression surgery has been shown to be highly successful with over 70% of individual's pain free after 10 years postoperatively (NHS[4]). One newer procedure is a stereotactic procedure referred to as gamma knife radiosurgery. Here a concentrated beam of radiation is directed towards the trigeminal nerve. It is a controversial area with emerging research demonstrating that patients treated multiple times with this surgery are more likely to get facial numbness than those who have had it a single time (Elaimy et al., 2012). Questions still remain about the efficacy of this procedure.

Dysfunction of the TMJ
Temporomandibular disorders are a collection of conditions affecting the TMJ. It is a relatively common condition affecting approximately 12% of the population but with many factors that contribute to it (Marklund and Wänman, 2007).

2.4.6 Abducent Nerve
The ciliary ganglion is the peripheral ganglion of the parasympathetic system of the eye. It is located towards the rear of the orbit, lateral to the optic nerve, medial to the lateral rectus muscle, and in front or lateral to the ophthalmic artery. It is rather small and is frequently connected with the nasociliary nerve by communicating branches

---

[4]http://www.nhs.uk/conditions/trigeminal-neuralgia/Pages/Introduction.aspx (Accessed September 28, 2015).

(sensory root). The short ciliary nerves are numerous branches of the ganglion that are distributed to the eyeball. Some of the afferent fibers from the eye travel by the short ciliary nerves through the ganglion and reach the nasociliary nerve. The fibers connected with the ganglion are described as its roots. A parasympathetic motor root or roots come from the branch of the oculomotor nerve to the inferior oblique. These fibers synapse in the ganglion and are the only fibers which do so. The postganglionic fibers pass to the short ciliary nerves and supply the ciliary muscle and the sphincter pupillae. The sympathetic fibers are derived from the internal carotid plexus and reach the ganglion either directly or by means of the nasociliary nerve, and a separate sympathetic root is seldom found. These fibers are postganglionic, arising in the superior cervical ganglion. They pass through the ciliary ganglion and, by the short ciliary nerves, innervate the dilator pupillae, palpebral or tarsal muscles, orbit and the blood vessels of the eyeball. Figs. 2.9 and 2.10 demonstrate the nerve and related structures.

Clinical Applications
As with all clinical examinations, introduce yourself to the patient, stating who you are and your purpose for meeting with the patient, especially if you are a student. Always take a comprehensive history before examining the patient. This will guide you to the most appropriate examinations that need to be conducted. Ensure you explain everything to the patient prior to doing it. This ensures a better level of trust, and excellent communication exists during the consultation.

Figure 2.9 An unlabeled image to demonstrate the position of the abducent nerve in relation to other structures.

Testing of the Abducent Nerve at the Bedside

Testing of the abducent nerve is undertaken when assessing the oculo-motor nerve and the trochlear nerves at the same time. It involves test-ing of all of the extraocular muscles at the same time. Therefore, the procedure for testing the abducent nerve is exactly the same as that for the oculomotor nerve and the trochlear nerves:

1. Ask the patient to keep their head still during the examination to ensure it is the extraocular muscles that are observed, rather than a false positive from movement of the head relative to the neck.
2. With their eyes only, the patient should follow the tip of your finger (or pen torch, or pencil etc.).
3. The examiner should then move the object in the horizontal plane from the extreme left to the extreme right.
4. This should be done *slowly.*
5. When at the extreme left or right, with the examining object, stop!
6. Observe for nystagmus.
7. Then an H-shaped pattern with the examining object can be drawn in space, which the patient should follow *with their eyes only.* This is to prevent any involvement of movement of the head leading to misinterpretation of the findings.
8. All movements should be done *slowly* to assess for eye movements.

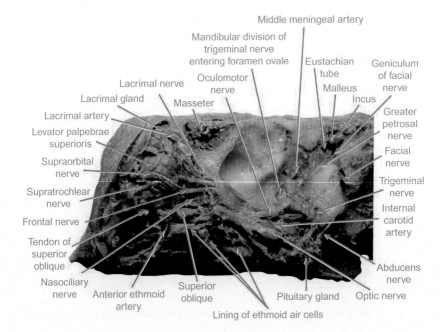

Figure 2.10 A labeled image to demonstrate the position of the abducent nerve in relation to other structures.

Advanced Testing

Advanced testing can be undertaken by a variety of means, generally with the input of a specialist, for example, ophthalmologist, and/or neurological input. Three broad categories of detailed examination can be done including the rotational movement of each eye (as previously described), comparison of yoke muscles (pairs of muscles that move the eyes in conjugate direction) and the red lens diplopia test (using diplopia as a test for weakness of the eye muscles) (Walker et al., 1990). These advanced tests, used with specialist input, can help establish pathology of all of the extraocular muscles and the nerves that supply them, that is, oculomotor, trochlear, and abducent nerves.

In addition, further testing will be directed by what is found on clinical examination. As the abducent nerve has a long course from its origin at the pontomedullary junction to its termination in the lateral rectus, there are many causes for compression of this nerve. Any cause of an increased intracranial pressure, aneurysm, space occupying lesion, cavernous sinus thrombosis, or atherosclerotic plaque within the internal carotid artery may give rise to pressure on the abducent nerve. Therefore, the presentation of the patient, detailed history, and clinical examination will direct the most appropriate investigations, in consultation with the relevant specialist.

Pathologies

Pathologies of the abducent nerve can be divided into those affecting the nucleus and the peripheral component of the nerve.

1. *Lesion of the abducent nerve nucleus*

   The abducent nerve nucleus contains two types of fibers in it: those that control the ipsilateral abducent nerve and interneurons that connect with the oculomotor nucleus of the opposite side by crossing the midline. This allows for controlled eye movement when lateral gaze of one eye is coupled with medial gaze of the contralateral eye (by the medial rectus of that side). This conjugate gaze is controlled by the medial longitudinal fasciculus. Lesions of this pathway and the abducent nerve results in *internuclear ophthalmoplegia*. When you ask the patient to look to the unaffected eye, the affected eye shows only minimal adduction. On inspection of the contralateral eye (to the site of pathology), that eye will abduct but will have nystagmus. The patient will complain of diplopia (double vision) on looking towards the *unaffected* eye. In younger patients with multiple sclerosis, bilateral internuclear ophthalmoplegia can be found.

There is a rare syndrome referred to as Wall-Eyed Bilateral INternuclear Ophthalmoplegia (WEBINO syndrome). Here the patient will have a bilateral exotropia on primary gaze, bilateral internuclear ophthalmoplegia with impaired convergence (Chakravarthi et al., 2013). Many causes have been identified for this pattern, but the most common is infarction at the level of the midbrain (Chen and Lin, 2007). If the lesion affects the abducent nerve nucleus or the paramedian pontine reticular formation as well as the medial longitudinal fasciculus on the same side, it will result in conjugate horizontal gaze palsy in one direction and internuclear ophthalmoplegia in the other. This is referred to as the one and a half syndrome, typically caused by multiple sclerosis, brainstem stroke, or tumor or arteriovenous malformations at the level of the brainstem (Wall and Wray, 1983).

2. *Lesions of the peripheral abducent nerve*

A variety of pathologies can cause an abducent nerve paralysis. Table 2.11 highlights some of the more common reasons for this.

2.4.7 Facial Nerve

The facial nerve, or seventh cranial nerve, has a complicated course after its origin from the pontomedullary junction. The facial nerve consists of a larger component which supplies the muscles of facial

Table 2.11 Summary Table Showing Typical Pathologies Which Can Affect the Abducent Nerve	
Causes of abducent nerve palsy	
Trauma	Head trauma e.g. base of skull fractures
Vascular	Aneurysm Cavernous sinus disease Infarction (stroke) Intra-cerebral haemorrhage
Neurological	Diabetic neuropathy (most common cause) Variety of neuropathies Demyelination e.g. multiple sclerosis
Infection	Meningitis
Neoplastic	Any space occupying lesion within the cranial cavity compressing on the abducent nerve
Congenital	An isolated abducent nerve pathology in a child should always be further investigated for an intracranial tumour as the cause
Iatrogenic	The abducent nerve is the most common nerve to be injured from insertion of a halo orthosis for cervical spinal injuries (Benzel, 2012).

expression, and a smaller part, which is referred to as the nervus intermedius. The smaller component, the nervus intermedius, contains the fibers for taste from the anterior two-thirds of the tongue and secretomotor fibers for the lacrimal and salivary glands (expect the parotid gland).

The two components of the facial nerve (main trunk and the nervus intermedius) leave the brain at the lower border of the pons (in the cerebellopontine angle) and together with the vestibulocochlear nerve, the eight cranial nerve, enter the internal auditory (acoustic) meatus. The facial nerve then courses laterally in the meatus and then enters the facial canal in the temporal bone. Above the promontory on the medial wall of the middle ear, the nerve is expanded to form the geniculate ganglion. The geniculate ganglion contains the cells of origin of its taste fibers. The facial nerve then turns sharply backwards, the bend called the geniculum (genu). The nerve then sweeps downward behind the middle ear and emerges from the skull at the stylomastoid foramen. The mastoid process is not well developed in infants and can place the nerve at risk of damage, especially forceps delivery, so this has to be kept in mind with the positioning of forceps during a delivery that requires some intervention from the midwifery team. The facial nerve then enters the posterior aspect of the parotid gland, divides into its cervicofacial and temporofacial divisions at the pes anserinus, and then terminates into its finer branches across the face.

In its course, the facial nerve traverses in succession:

1. the posterior cranial fossa
2. the internal auditory (acoustic) meatus
3. the facial canal in the temporal bone and
4. the parotid gland and the face.

There are several branches of thee facial nerve and can be summarized as follows:

1. *Larger component*
   a. *motor* fibers
   b. supplies the *muscles of facial expression* (most important clinically!)

2. *Smaller component*
   a. the *nervus intermedius* (or intermediate nerve as it is sandwiched between the facial and vestibulocochlear nerves)
   b. *sensory* and *parasympathetic* fibers
   c. supplies the *anterior 2/3 of the tongue* (taste fibers) and *lacrimal, submandibular. and sublingual salivary glands* (secretomotor).

Now, a description of the branches of the facial nerve will be given.

Greater Petrosal Nerve

In the internal acoustic meatus, the facial nerve communicates with the vestibulocochlear nerve. In the facial canal, the geniculate ganglion gives rise to several branches. The first of these is the greater petrosal nerve. The greater petrosal nerve passes forward in a groove towards the foramen lacerum. There it is joined by the deep petrosal nerve (from the sympathetic plexus on the internal carotid artery) to form the nerve of the pterygoid canal and reaches the pterygopalatine ganglion. The greater petrosal nerve contains secretomotor fibers for the lacrimal and nasal glands, and perhaps vasodilator fibers for the middle meningeal arteries. It also contains a number of afferent fibers (the cells of origin being in the geniculate ganglion); their distribution and function are uncertain, but some of the fibers may subserve general sensation from the nasal mucosa.

Nerve to Stapedius

The stapedius is the smallest skeletal muscle in the body and is approximately 1 mm in length. It arises from a prominence in the tympanic cavity at the posterior aspect called the pyramidal eminence. It inserts into the neck of the stapes. The nerve to stapedius arises opposite the pyramidal eminence from the facial nerve, and it passes through this canal to innervate the stapedius.

Tip!

Injury of the facial nerve distal to the geniculate ganglion before its branch to the stapedius will result in patient's experiencing louder noises on the affected side. This is because stapedius normally functions to prevent excessive movements of the stapedius thus controlling the amplitude of the sound waves passing to the inner ear. If there is damage to the nerve to stapedius, wider oscillations of the stapes will occur resulting in hyperacusis—sounds being perceived as extremely loud, more so than they actually are to a person without damage to this nerve.

Chorda Tympani

The chorda tympani enters the tympanic cavity (although covered by a reflection of the mucous membrane), passes medial to the tympanic membrane and the handle of the malleus, and again enters the temporal bone. It exits the skull through the petrotympanic fissure and descends in the infratemporal fossa. Medial to the lateral pterygoid muscle, it joins with the lingual nerve and is then distributed to the anterior two-thirds of the side and dorsum of the tongue. The chorda tympani contains several types of fibers:

1. Those associated with taste and common sensation (including pain) from the anterior two-thirds of the tongue and from the soft palate
2. Preganglionic secretory and vasodilator fibers, which synapse in the submandibular ganglion, the postganglionic fibers then supplying the submandibular, sublingual, and lingual glands.

Below the base of the skull, the chorda tympani communicates with the otic ganglion (see "Mandibular Nerve").

On exiting the skull at the stylomastoid foramen, the facial nerve then enters the parotid gland and starts dividing at the pes anserinus (goose's foot). The following branches then are the final branches from the main trunk of the facial nerve:

Temporal Nerve

The temporal branches of the facial nerve reach the temporal region by crossing the zygomatic arch. They innervate auricular muscles like the superior and anterior auricular muscles, and anastomose with the zygiomaticotemporal branch form the maxillary nerve and also the auriculotemporal nerve. Anteriorly, the temporal branches also innervate the orbicularis oculi, frontal belly of the occipitofrontalis, and the corrugator. These branches then anastomose with the lacrimal and supraorbital branches from the ophthalmic nerve.

Zygomatic Nerve

The zygomatic branches of the facial nerve pass across the zygomatic bone innervating the orbicularis oculi and anastomose with the lacrimal nerve and also the zygomaticofacial branch from the maxillary nerve.

Buccal Nerve

The buccal branches of the facial nerve run horizontally to pass below the orbit and around the mouth. The buccal nerve has a superficial and deep branch. The superficial branch passes sandwiched between the skin and the superficial muscles of the face. It innervates the muscles at this site and anastomoses with the external nasal nerve and the infratrochlear nerves. The deep branches pass under the levator labii superioris and the zygomaticus major innervating them but also forming an infraorbital plexus with the superior labial branches which arise from the infraorbital nerve. The deep branches also innervate the zygomaticus minor, levator anguli oris, and the levator labii superioris alaeque nasi.

Marginal Mandibular Nerve
The marginal mandibular branches of the facial nerve dips below the angle of the mandible, then rises above the level of the body of the mandible to innervate the muscles of the chin, for example, depressor anguli oris and depressor labii inferioris.

Cervical Nerve
The cervical branch of the facial nerve innervates the platysma approaching it from its deep aspect.

Posterior Auricular Nerve
The posterior auricular nerve accompanies the posterior auricular artery and supplies the muscles of the auricle together with the occipitalis. The superior and anterior auricular muscles however are innervated by the temporal branches of the facial nerve. The posterior auricular nerve also supplies sensory fibers to the auricle.

Tip!
Testing of the posterior auricular nerve could involve testing for general sensation and pain in the auricle or by trying to assess the function of occipitalis. However, it is seldom undertaken in the clinical setting. If there is an issue with the facial nerve and the structures it supplies, there may well be other branches of the facial nerve that are required to be examined, for example, those supplying the muscles of facial expression. A full account of clinical examination of the facial nerve is given later.

Branch to Posterior Belly of Digastric
The branch to the posterior belly of digastric is also referred to as the digastric branch. This branch comes off the facial nerve just after the facial nerve exits the stylomastoid foramen. The branches then innervate the posterior belly of digastric. As well as this, a small branch communicates with the glossopharyngeal nerve.

Stylohyoid Branch
The stylohyoid muscle is a slender muscle found along the upper border of the posterior belly of digastric. It arises from the back of the styloid process and is inserted into the hyoid bone at the junction between the body and the greater horn. It is typically split near its insertion by the tendon of the digastric. The purpose of this muscle is that it draws the hyoid bone backwards and elongates the floor of the mouth. The anteroposterior position of the hyoid bone is determined by the stylohyoid, geniohyoid, and the infrahyoid muscles.

The stylohyoid branch to this muscle, forms the facial nerve, originates close to the stylomastoid foramen, dividing into several smaller filaments, which innervates the muscle. In addition to this, communicating branches pass to the glossopharyngeal and vagus nerves, the auricular branch of the vagus, and the auriculotemporal, great auricular, and lesser occipital nerves.

Clinical Applications
As with all clinical examinations, introduce yourself to the patient, stating who you are and your purpose for meeting with the patient, especially if you are a student. Always take a comprehensive history before examining the patient. This will guide you to the most appropriate examinations that need to be conducted. Ensure you explain everything to the patient prior to doing it. This ensures a better level of trust, and excellent communication exists during the consultation.

The facial nerve is tested chiefly in regard to the facial muscles as follows:

Testing of the Facial Nerve at the Bedside
1. The purpose of the facial nerve is to ensure functioning of the muscles of facial expression
2. Inspect the face during conversation and rest noting any asymmetry (eg, drooping, sagging and even smoothing of the normal facial creases)
3. Then do the following:
   a. Ask the patient to raise their eyebrows, then
   b. ask the patient to frown, then
   c. ask the patient to show you their teeth.

Tip!

Don't ask them to "smile" as they may be very worried about their signs and symptoms/clinical condition, and may feel uncomfortable being asked to smile! On the other hand, make sure they have their own teeth/substitutes in place to prevent embarrassment.

   d. Puff out their cheeks against pursed lips.
   e. Scrunch up their eyes, and as the examiner, try to open them on behalf of the patient.

Tip!

Do tell them what you are about to do, as the patient will feel surprised that you are trying to prise their eyes open!

4. The purpose of the examination is to note asymmetry, but also, to determine the strength (or weakness) of the power of the facial muscles.

Tip!

Always test the resistance of the muscles of facial expression of the patient, as the patient may try and disguise their problem as they may be in denial of their underlying condition.

Advanced Testing

Taste can be tested on the anterior 2/3 of the tongue with a swab dipped in a flavored solution to test sweet/salt/sour and/or bitter substances on one half of the protruded tongue on the side to be formally examined. Lacrimation is rarely formally tested for.

Tip!

Always get the patient to rinse their mouth out after testing each of the standard tastes. Typically, this involves using sweet or salt using salt or sugar, respectively. The patient should be told what you would do before doing it. A small tongue blade can be used to place a small amount of salt or sugar onto each part of the tongue needed testing. It should be placed on the lateral aspect of the tongue. Remember when testing the facial nerve for salt or sugar taste sensation, it only supplies the anterior two-thirds.

The corneal reflex may be performed where a blink reflex is initiated in the patient by *gently* touching the corneal surface with a soft object, for example, cotton wool. The afferent (going to the brain) arc is with the trigeminal nerve (ophthalmic division, see Trigeminal Nerve chapter). The efferent component is via the facial nerve, so this test actually tests two of the cranial nerves at the same time—five (trigeminal) and seven (facial).

The Ganglia of the Facial Nerve

Table 2.12 summarizes the ganglia associated with the facial nerve, including its location and function(s).

Pathologies

Many types of pathologies affect the facial nerve, with Bell's palsy being the most frequent, and of all cranial nerves to be affected by pathology, it is the facial nerve with this condition. Also, the site of the pathology involving the facial nerve can be determined by the signs and symptoms presented by the patient. This can help localize the anatomical site to aid investigation of the patient. Table 2.13 summarizes the types of presentation of a facial nerve pathology, and where the lesion may well be at. It can help determine the anatomical site of such a lesion.

Facial paralysis is not an uncommon clinical condition with numerous etiologies. From a clinical perspective, it is most important to

Table 2.12 Summary Table of the Facial Nerve Ganglia, Where They Are Found Anatomically, and the Functions Related to Each One		
Ganglion	Location	Function
Geniculate	Facial canal (petrous temporal bone)	Sensory ganglion Special sensory neuronal cell bodies for taste Fibers from motor, sensory and parasympathetic functions pass through here
Pterygopalatine	Pterygopalatine fossa	Synapses here for pre- and post-synaptic parasympathetic innervation of the lacrimal glands Sensory and sympathetic fibers also pass through it
Submandibular	Above the deep portion of the submandibular gland on hyoglossus	Synapses here for pre- and post-synaptic parasympathetic innervation of the submandibular and sublingual salivary glands

Table 2.13 Types of Presentations a Patient May Have With a Variety of Pathologies Affecting the Facial Nerve

Site of Pathology	Branches affected	Type of Pathology	Clinical Presentation
Proximal to the geniculate ganglion	Main trunk of the facial nerve Nerve to stapedius Chorda tympani	Acoustic neuroma (vestibular Schwannoma) Other intracranial tumor	All functions of the nerve affected: Facial paralysis Hyperacusis Impaired secretion of tears Loss or impairment of taste on the anterior 2/3 of the tongue Loss of taste on the palate on the affected side Loss of salivary secretions from the submandibular and sublingual glands
Upper part of facial canal	Main trunk of the facial nerve Chorda tympani	Infections Bell's palsy	Facial paralysis Loss or impairment of taste on the anterior 2/3 of the tongue Loss of salivary secretions from the submandibular and sublingual glands
Lower part of facial canal	Main trunk of the facial nerve	Infections Bell's palsy	Facial paralysis
Parotid gland	Main trunk of the facial nerve or its individual branches (e.g. temporal, zygomatic, buccal, marginal mandibular or cervical)	Parotid tumor Parotitis (e.g. mumps) Direct trauma Parotidectomy	Complete or partial facial paralysis dependent on the branch(es) involved

differentiate between an upper motor neuron pathology and a lower motor neuron pathology.

1. *Upper motor neuron pathology*
   The commonest cause of an upper motor neuron pathology is a stroke (cerebrovascular disease), though other causes can include intracranial tumors, infections (eg, HIV, syphilis) or vasculitic diseases.
   A stroke results in interruption of the fibers from the motor regions of the cerebral cortex as they pass through the internal capsule to the facial motor nucleus. This causes a voluntary paralysis of the muscles of facial expression by the facial nerve on the opposite side of the lesion. However, the upper part of the facial motor nucleus receives both crossed and uncrossed fibers resulting in the orbicularis oculi and frontalis being spared to varying degrees.

KEY POINT—If a patient has had a stroke involving the supra-nuclear fibers to the facial nucleus, they are still able to wrinkle their forehead and close the eye on the contralateral side. This is despite the marked paralysis of the rest of the face including a droopy corner of the mouth and puffing of the cheek.

2. *Lower motor neuron pathology*

In a lower motor neuron paralysis involving the facial nerve, all muscles on the side of the pathology are affected, including frontalis and the orbicularis oculi. Therefore, these patients are unable to wrinkle their forehead or close their eye on the affected side. The most common cause for a lower motor neuron pathology involving the facial nerve is Bell's palsy, named after the famous Scottish anatomist and surgeon Sir Charles Bell (1774—1842) who first described it. The other causes of a lower motor neuron pathology involving the facial nerve are summarized below.

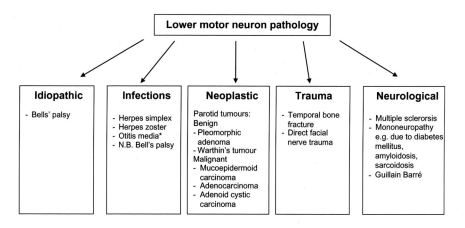

Today, Bell's palsy, the commonest cause of a facial paralysis, is understood to involve a viral infection (many have been implicated). The infection then results in inflammation and compression within the facial canal. As this canal is very small and no room exists for expansion, the facial nerve is compressed due to edema of the tissues lining the canal pressing on the facial nerve. This reduces or blocks the nervous impulse transmission to the muscles of facial expression resulting in weakness or paralysis of those muscles.

Dependent on the site of pathology determines the clinical presentation of the patient. Below details a summary table of what the patient may present with, taking into account the anatomical components of the facial nerve that may be affected.

2.4.8 Vestibulocochlear Nerve

The vestibulocochlear nerve arises from the pontomedullary junction behind the facial nerve. Although there are two different central connections for the vestibulocochlear nerve, dependent on what component is being served, the main function is in the transmission of afferent information from the inner ear to the brain.

The vestibular ganglion is associated with the vestibular nerve and the spiral ganglion with the cochlear nerve. Both of these ganglia associated with the vestibulocochlear nerve are bipolar cells where the central portion goes to the brain and the peripheral portion goes to the inner ear.

The vestibulocochlear nerve passes through the posterior cranial fossa to enter the petrous temporal bone at the internal auditory (acoustic) meatus. Within the internal auditory meatus, the vestibulocochlear nerve is accompanied by the two branches of the facial nerve (motor nerve and nervus intermedius and the labyrinthine vessels. At this point, numerous accounts have been given in relation to communication between the vestibulocochlear and adjacent facial nerves (Paturet, 1951; Fisch, 1973; Shoja et al., 2014). Towards the lateral end of the internal auditory meatus, it divides into the more anterior cochlear nerve and the more posterior vestibular nerve supplies the saccule and utricle, as well as the ampullary crests of the semicircular ducts.

Within the petrous temporal bone, the inner ear is located. The inner ear has two functions—hearing and balance (equilibration). The inner ear comprised the membranous labyrinth within the bony labyrinth. Sensory information for the maintenance of equilibrium comes from three systems—the eyes, proprioceptive endings throughout the entire body, and the vestibular apparatus of the inner ear. The static labyrinth, comprised the utricle and saccule (endolymph dilations of the membranous labyrinth), detects the position of the head with respect to gravity. The kinetic labyrinth detects movement of the head via the three semicircular canals.

For hearing, the cochlear part of the inner ear contains the organ of Corti, also referred to as the spiral organ. Sound waves are transmitted to the fenestra vestibule (also known as the oval window) via the vibrations of the stapes of the middle ear. The fenestra cochleae (also known as the round window) are below the oval window. It is closed by a thin membrane which allows for pressure waves to be generated in the inner ear; otherwise it would just be a fixed, rigid box with no movement.

The cochlea makes two-and-a-half turns around the core called the modiolus. Within the cochlea, there is a cochlear duct (scala media) that contains endolymph and is firmly fixed to the inner and outer walls of the canal. In close communication with the fenestra vestibule, there is the perilymph in the scala vestibule. Communicating directly with this is the perilymph found in the scala tympani, closely related to the fenestra cochleae.

As there are two distinct functions dealt with by the vestibulocochlear nerve, it has two separate nerves: the vestibular and cochlear nerves which will now be dealt with in turn. Refer to Figs. 2.11 and 2.12 which demonstrate this nerve and related structures.

Vestibular Nerve
The vestibular ganglion is found in the trunk of the nerve at, or within, the internal auditory meatus. On the more distal side of the vestibular ganglion the nerve divides into a posterior, superior, and inferior branch. The fibers are distributed as follows:

a. *Posterior division*—passes to the ampullary crest of the posterior semicircular duct.
b. *Superior division*—passes to the macula of the utricle and the ampullary crests of the lateral and anterior semicircular ducts.
c. *Inferior division*—passes to the macula of the saccule.

Cochlear Nerve
The spiral ganglion is found at the spiral canal of the modiolus. The fibers from here pass to the edge of the osseous spiral lamina. Some fibers pass to the outer hair cells, while most passes to the basal and middle coils.

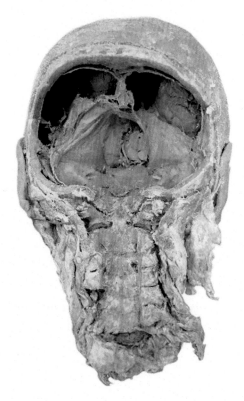

Figure 2.11 An unlabeled image to demonstrate the position of the vestibulocochlear nerve in relation to other structures.

From the spiral ganglion, two cell types have been identified (Spoendlin, 1988). These are described as follows:

Type I—Large bipolar myelinated cells with long axons projecting centrally and peripherally. Type I cells account for the majority of cochlear nerve cells and are for the inner hair cells. Ten ganglion cells are connected to each sensory cell.

Type II—Small nonmyelinated cells with a peripherally directed process. These are smaller in number (approximately 10%) and are afferent for the outer hair cells.

Therefore, the funnel-shaped external auditory canal collects sound. The purpose of this structure is to amplify the sound transmitting it to the tympanic membrane. Then, the malleus, incus, and stapes amplify this signal and allow the sound transmission to be converted to a mechanical effect due to the vibration of the tympanic membrane.

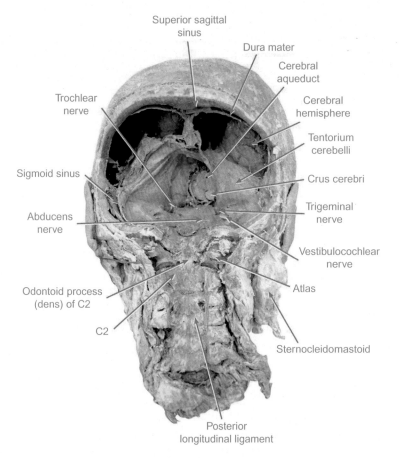

Figure 2.12 A labeled image to demonstrate the position of the vestibulocochlear nerve in relation to other structures.

The middle ear also dampens any excessive vibration and therefore allows for impedance matching. The force at the stapes per unit area of oscillating surface is increased some 20-fold (Williams and Warwick, 1980). The basilar membrane of the inner ear then receives this information and is involved in mechanical and neural filtering and analysis of signals by the spiral organ. The inner ear and basilar membrane are also involved in stimulus transduction and initiation of the action potentials between the cochlear nerve cells and the sensory neurons. This then allows for the conversion of sound into something electrical, and that the higher brain functions can become involved in interpreting.

Clinical Applications
Vestibular Nerve
As with all clinical examinations, introduce yourself to the patient, stating who you are and your purpose for meeting with the patient, especially if you are a student. Always take a comprehensive history before examining the patient. This will guide you to the most appropriate examinations that need to be conducted. Ensure you explain everything to the patient prior to doing it. This ensures a better level of trust, and excellent communication exists during the consultation.

Testing of the vestibulocochlear nerve at the bedside:

1. *Basic testing*
   Assess hearing by instructing the patient to close their eyes and to say "left" or "right" when a sound is heard in the respective ear. Vigorously rub your fingers together very near to, yet not touching, each ear and wait for the patient to respond. After this test, ask the patient if the sound was the same in both ears or louder in a specific ear. If there is lateralization or hearing abnormalities perform the Rinne and Weber tests using the 512 Hz tuning fork.
2. *Weber test*
   The Weber test is a test for lateralization. Tap the tuning fork strongly on your palm and then press the butt of the instrument on the top of the patient's head in the midline and ask the patient where they hear the sound. Normally, the sound is heard in the center of the head or equally in both ears. If there is a conductive hearing loss, the vibration will be louder on the side with the conductive hearing loss. If the patient does not hear the vibration at all, attempt again, but press the butt harder on the patient's head.
3. *Rinne test*
   The Rinne test compares air conduction to bone conduction. Tap the tuning fork firmly on your palm and place the butt on the mastoid eminence firmly. Tell the patient to say "now" when they can no longer hear the vibration. When the patient says "now," remove the butt from the mastoid process and place the U of the tuning fork near the ear without touching it.
   Tell the patient to say "now" when they can no longer hear anything. Normally, one will have greater air conduction than bone conduction and therefore hear the vibration longer with the fork in the air. If the bone conduction is the same or greater than the air

conduction, there is a conductive hearing impairment on that side. If there is a sensorineural hearing loss, then the vibration is heard substantially longer than usual in the air.

Make certain that you perform both the Weber and Rinne tests on both ears. It would also be prudent to perform an otoscopic examination of both eardrums to rule out a severe otitis media, perforation of the tympanic membrane, or even occlusion of the external auditory meatus, which all may confuse the results of these tests. If hearing loss is noted, an audiogram is indicated to provide a baseline of hearing for future reference.

4. *Otoscopy*

   This will allow the external auditory meatus and the middle ear to be assessed by examination of the tympanic membrane.

Advanced Testing of the Vestibulocochlear Nerve

1. *Audiometry (hearing) tests*

   a. *Automated otoacoustic emission (AOAE) test*

   The AOAE test works on the basis that the normal cochlea, when stimulated by sound, produces an echo. Therefore, this test involves an earpiece with a speaker and microphone connected. Clicking sounds are played through the speaker, and if the cochlea is functioning, the microphone will detect the echo.

   This is a simple and effective method to quickly assess the hearing of a patient, including that of the very young patient. It allows for an immediate result to be given, and from this it can then be classified as "pass" or "refer." "Refer" does not necessarily mean that there is a gross abnormality. It may simply be that there was too much background noise in the room interfering with the test, or in the very young patient, fluid from childbirth may still have been present in the external auditory meatus, or the child may have just been too restless. More formal testing can therefore help to give a clearer indication of the patient's hearing status.

   b. *Automated auditory brainstem response (AABR) test*

   Sensors are placed on the patient's head and neck. It works by sounds being played through headphones which the patient wears for the duration of the test. The sensors will detect these sounds being transmitted. A small percentage of patient's (generally babies) can be referred on for a more in-depth analysis of their hearing responses.

c. *Pure tone audiometry (PTA) test*

This test is one of the key hearing tests, but only suitable for adults and children old enough to be able to communicate effectively and cooperate with the test. This test tests both air and bone conduction. In a soundproof room, the patient has headphones placed on, as well as a headband to assess bone conduction. A variety of frequencies and volumes are then played, and the patient then has to press a button on a machine to indicate when they hear the noise. The headband can assess the bone conduction.

d. *Bone conduction tests*

This test is a more advanced form of using tuning forks and combined with a pure tone audiometry test, allows differentiation of external, middle, or inner ear pathology.

A variety of other formal tests are also available in assessing hearing including tympanometry, electrophysiological, and acoustic reflex testing. For specialist advice, it is wise to consult your local audiologist and/or otorhinolaryngologist.

2. *Vestibular tests*

a. *Rotation testing*

This test assesses eye movements when the head is rotating at a variety of speeds. Two types exist—auto head rotation and rotary chair. With the former, the patient moves their head back and forward observing a fixed target. With the rotary chair, a computer will control the movement of the chair, and the eye movements are recorded.

b. *Electronystagmography (ENG)*

This test assesses nystagmus with electrodes placed on the skin around the eye. The other method for assessing this is to use videonystagmography where an infrared camera in glasses monitors eye movements. Overall, the test observes the eye either in a fixed position in different directions or having the eyes follow a moving target.

c. *Computerized dynamic posturography*

This test assesses postural stability by standing on a platform with a visual target to observe. The visual target and/or the platform will move and the three inputs for posture are therefore assessed, that is, visual, vestibular, and sensory information from the joints and muscles.

d. *Vestibular evoked myogenic potential*

By placing headphones over the ears and electrodes on the skin overlying the neck musculature, a sound impulse is played. The purpose of this study is to assess the response of vestibular stimuli and assesses the inferior vestibular nerve and the saccule.

e. A variety of hearing tests can be undertaken as described in the previous section.

f. As directed by the patient's history and examination, other tests could be an MRI, which will aid in the diagnosis of any soft tissue problem, and/or CT which will help in the diagnosis perhaps of temporal bone pathology.

Pathologies

A number of pathologies can affect the vestibulocochlear nerve as detailed below.

1. *Deafness*

Two different types of deafness exist—conductive and sensorineural.

a. *Conductive deafness*

With conductive deafness, there is impaired transmission of the sound waves along the external auditory meatus to the middle ear (stapes). A variety of causes can result in conductive deafness like ear wax, discharge from otitis externa, otitis media (supportive and serous), and damage to the tympanic membrane (perhaps from rupture or scarring), foreign body, or congenital abnormality.

b. *Sensorineural deafness*

Sensorineural deafness is caused by pathology central to the oval window in the cochlea (therefore sensory), cochlear part of the vestibulocochlear nerve (therefore neural), or perhaps within central pathways. The most common cause of this is exposure to loud noises, especially over prolonged periods of time. Other notable causes of sensorineural hearing loss are infections/inflammation, trauma (base of skull fracture), ischemia, and ototoxic drugs, for example, gentamycin.

2. *Vertigo*

Vertigo is a feeling of disorientation and can include feeling dizzy, or that the environment is moving or spinning. Nausea and vomiting, nystagmus, and hearing loss can accompany it. The causes of vertigo are described as either being central or peripheral in origin. Central causes can include neurological (multiple sclerosis), vascular (vertebrobasilar ischemia), infectious (syphilis, herpes), trauma to the head, or tumor (acoustic neuroma). Peripheral causes can include vestibular neuronitis, cholesteatoma, labyrinthitis, benign postural vertigo or Ménière's disease.

3. *Tinnitus*

   Tinnitus is where the patient will experience ringing or buzzing in the ears. The exact mechanism is uncertain; however, related causes can be trauma, drugs (eg, aspirin, loop diuretics, gentamycin), infection (suppurative otitis media, viral), psychological issues.

4. *Acoustic neuroma*

   An acoustic neuroma is also called a vestibular Schwannoma. This tumor arises from the Schwann cells. It is a slow growing tumor, growing at approximately 1−2 mm per year. As it is a slow growing tumor, it takes some time before signs and symptoms present. When they do present, hearing loss will be the most obvious symptom. It is a sensorineural deafness and can also present as vertigo, nausea, altered balance, and the majority of patient's will have tinnitus. The cause of an acoustic neuroma is generally unknown but can be related to von Recklinghausen neurofibromatosis.

2.4.9 Glossopharyngeal Nerve

The glossopharyngeal nerve is the ninth cranial nerve. It is rather complex having four related nuclei conveying information related to sensation, muscular activity, and autonomic function. It contains *special visceral efferent (visceral motor), general visceral efferent (parasympathetic), general visceral afferent (visceral sensory),* and *general somatic afferent (general sensory)* information from a variety of structures.

The glossopharyngeal nerve contains sensory fibers from the pharynx, tongue (posterior one-third), and the tonsils. It also contains secretomotor fibers destined for the parotid gland as well as motor fibers for the stylopharyngeus. Finally, it also contains taste fibers, also from the posterior one-third of the tongue. Therefore, the glossopharyngeal nerve is a rather complex and important nerve supplying a variety of structures.

The glossopharyngeal nerve arises as three or four rootlets at the level of the medulla oblongata. It passes out from between the inferior cerebellar peduncle and the olive, superior to the rootlets of the vagus nerve. It then sits on the jugular tubercle of the occipital bone. It then runs to the jugular foramen, passing through the middle part of it. At the point of entry to the jugular foramen, two ganglia are found— an inferior and superior ganglion. Both of these ganglia contain the cell bodies of the afferent fibers contained within the glossopharyngeal nerve. On passing through the jugular foramen, the glossopharyngeal nerve then passes between the internal carotid artery and the internal jugular vein, descending in front of the artery. It then passes deep to the styloid process and related muscles attaching on to this bony prominence. It then winds round the stylopharyngeus, passing deep to the hyoglossus and going between the superior and middle pharyngeal constrictors. Figs. 2.13 and 2.14 demonstrate this nerve and related structures.

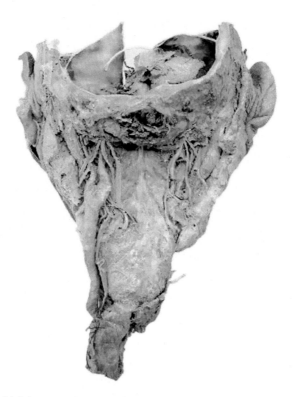

Figure 2.13 An unlabeled image to demonstrate the position of the glossopharyngeal nerve in relation to other structures.

The components of the glossopharyngeal nerve can be divided up as follows:

Special Visceral Efferent

The glossopharyngeal nerve contains branchial motor fibers to the sty-lopharyngeus muscle, which is derived from the third pharyngeal arch. This branch is given off to the muscle as it passes across it as it descends down from the jugular foramen. The stylopharyngeus muscle is responsible for raising the larynx and pharynx, and functions during swallowing.

General Visceral Efferent (Autonomic—Parasympathetic)

The general visceral efferent or parasympathetic fibers of the glosso-pharyngeal nerve initially arise from the tympanic nerve. The following

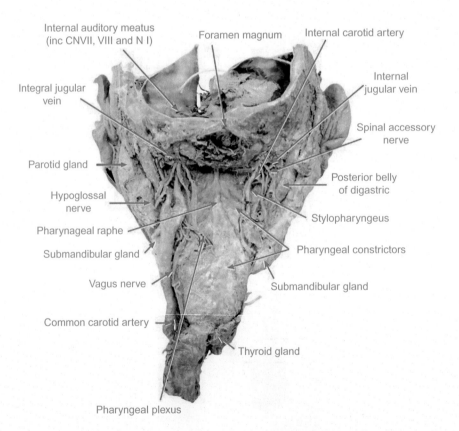

Figure 2.14 A labeled image to demonstrate the position of the glossopharyngeal nerve in relation to other structures.

sequence then occurs for the tympanic nerve to supply the parotid gland, as detailed below.

1. The tympanic nerve arises from the *inferior ganglion.*
2. The tympanic nerve passes to the tympanic cavity, passing through the *tympanic canaliculus.*
3. The tympanic nerve divides into branches forming the *tympanic plexus.*
4. From the tympanic plexus, two branching patterns arise: one to pass to the mucous membranes of the tympanic cavity, auditory tube, and mastoid air cells. The other branch gives the *lesser petrosal nerve.* It is this nerve that contains the fibers destined for the parotid gland.
5. The lesser petrosal nerve then passes through the temporal bone and then through the *foramen ovale.*
6. The lesser petrosal nerve then synapses in the *otic ganglion.*
7. Postsynaptic fibers then pass to the *auriculotemporal nerve*, a branch of the mandibular division of the trigeminal nerve
8. These parasympathetic fibers then pass to the parotid gland providing secretomotor fibers to it

General Visceral Afferent

The general visceral afferent, or general sensory fibers, has five distributions as highlighted in Table 2.14.

Special Visceral Afferent

The special visceral afferent fibers contained in the glossopharyngeal nerve are concerned with taste sensation from the posterior one-third of the tongue. The sensory information passes from the posterior

Table 2.14 The Distribution of the Glossopharyngeal Nerve Branches

Branch	Distribution
Carotid sinus branch	This branch passes to the anterolateral aspect of the internal carotid artery. It passes to the baroreceptors and chemoreceptors in the carotid sinus and the carotid body
Tympanic nerve	The sensory fiber component conveys information from the middle ear
Lingual branch	Conveys sensory information from the posterior one-third of the tongue as well as the vallate papillae
Tonsillar branch	Conveys information from the mucous membranes of the palatine tonsil and soft palate
Pharyngeal branch	This branch conveys sensory information from the oropharynx

one-third of the tongue to the pharyngeal branches of the glossopharyngeal nerve to pass to the inferior ganglion. From here, the fibers pass to the nucleus solitarius in the medulla where they synapse. From here, the fibers pass bilaterally to the thalamus via the ventral posteromedial nuclei and then onwards to the gustatory cortex within the parietal lobe.

General Somatic Afferent
The general somatic afferent or general sensory fibers convey general sensory information from the skin of the external ear, inside of the tympanic membrane, the upper portion of the pharynx, and general sensation from the posterior one-third of the tongue.

The fibers from the skin of the external ear initially travel with the vagus nerve (auricular branch (Arnold's nerve)). From the inner aspect of the tympanic membrane, the fibers for general sensation travel in the tympanic nerve (see above). The fibers for general sensation from the upper part of the pharynx and posterior one-third of the tongue pass via the pharyngeal branch of the glossopharyngeal nerve.

From these branches, they then pass centrally to the medulla entering the spinal nucleus of the trigeminal nerve, projecting contralaterally to the ventral posteromedial nucleus of the thalamus. From there, the fibers from the external ear, tympanic membrane, pharynx, and tongue then terminate in the sensory cortex for interpretation and processing of the information conveyed.

The ganglia of the glossopharyngeal nerve can be summarized as follows:

Superior ganglion—This ganglion is very small and is sometimes viewed as a broken off part of the inferior ganglion. It is found within the groove of where the glossopharyngeal nerve passes in the jugular foramen. It contains the visceral sensory fibers from the pharynx, parotid gland, carotid body, and sinus as well as the middle ear.

Inferior ganglion—The inferior ganglion conveys information related to special and general sensation from the mucous membrane of the posterior one-third of the tongue. Its peripheral fibers also come from the oropharynx and soft palate conveying general sensory fibers. The inferior ganglion is the bigger of the two ganglia related to the glossopharyngeal nerve and is found on the lower border of the petrous temporal bone in a notch. The inferior ganglion also communicates with the facial and vagus nerves, as well as with the sympathetic trunk.

The Important Branches of the Glossopharyngeal Nerve
Table 2.15 highlights the important branches of the glossopharyngeal nerve and what the functions of each of these branches are responsible for.

Tympanic Nerve
The glossopharyngeal nerve gives off the tympanic nerve. This nerve contains the parasympathetic fibers to parotid gland and general sensation to middle ear.

Carotid Branch
The glossopharyngeal nerve gives off the carotid branch. This nerve contains the visceral sensory fibers from the carotid body and sinus.

Pharyngeal Branches
The glossopharyngeal nerve gives off pharyngeal branches. These branches contain the special sensory fibers to the posterior one-third of the tongue. The pharyngeal branches also carry innervation for general sensation of the oropharynx.

Lingual Branches
The glossopharyngeal nerve gives off lingual branches. These lingual branches contain the general sensory fibers from the tongue (posterior one-third) and vallate papillae. Within the lingual branches, there may also be taste fibers from the posterior one-third of the tongue.

Table 2.15 Branches of the Glossopharyngeal Nerve and Their Functions	
Branch	Function
Muscular	Motor to stylopharyngeus
Tympanic	Parasympathetic fibres to parotid gland and general sensation to middle ear
Lingual	General sensation from the tongue (posterior one-third) and vallate papillae. May contain taste fibres from the posterior one-third of the tongue
Pharyngeal	Special sensory fibres to the posterior one-third of the tongue. General sensation to the oropharynx
Tonsillar	General sensation from the soft palate and palatine tonsil
Carotid sinus	Visceral sensory fibres from the carotid body and sinus

Tonsillar Branches

The glossopharyngeal nerve gives off tonsillar branches. These branches contain the general sensory fibers from the soft palate and palatine tonsil.

All of these branches, and what they innervate, are summarized in Table 2.15.

Clinical Applications

Testing of the Glossopharyngeal Nerve at the Bedside

1. The first thing to say is that examination of the glossopharyngeal nerve is difficult. Assessing it on its own is not possible, and an isolated lesion of this nerve is almost unknown (Walker et al., 1990). When assessing the glossopharyngeal nerve, the first thing to do is simply *listening to the patient talking.* Any abnormality of the voice, for example, hoarse, whispering, or a nasal voice may give a clue as to an abnormality. Also, ask the patient if they have any difficulty in swallowing. The result of a glossopharyngeal nerve (and related cranial nerves, eg, vagus and accessory nerves due to their close proximity to each other) may be dysphagia (difficulty swallowing), aspiration pneumonia, or dysarthria (difficulty in the motor control of speech).
2. To assess the function of the glossopharyngeal nerve (and the vagus nerve) ask the patient to say "ahhhh" (without protruding their tongue) for as long as they can. Normally, the palate should rise equally in the midline. The palate (uvula) will move *towards* the side of the lesion if there is a problem with the glossopharyngeal (and perhaps vagus) nerve.
3. Damage to the glossopharyngeal (and vagus) nerve, for example, form a stroke, may result in loss of the gag reflex. *Always* tell the patient what you will do before assessing the gag reflex, as it is not a pleasant examination, and may not always be necessary.
4. A swab can be used to gently touch the palatal arch on the left then right hand sides. Try to assess the normal side first if you suspect a pathology.

Tip!

If it is difficult to view the palate (and uvula) when examining the glossopharyngeal (and vagus) nerve, a tongue depressor allows for easier visualization. Again, always tell the patient what you will be doing before doing it, as some patient's do not like the taste/texture of a tongue depressor.

Advanced Testing of the Glossopharyngeal Nerve
It may be that taste from the posterior one-third of the tongue will need to be tested though is not commonly performed. This should be done in the same way that the facial nerve is tested for taste. *Always tell the patient what you will be doing before examining them.* This allows the patient to be fully informed and consent to the procedure but also will build up trust between you and the patient.

A small tongue blade can be used to place a small amount of salt or sugar onto each part of the tongue needed testing. It should be placed on the lateral aspect of the tongue. Remember that the glossopharyngeal nerve only supplies the posterior one-third of the tongue for taste sensation.

Pathologies
Isolated glossopharyngeal nerve palsy is extremely rare. Indeed, unilateral lesions of the glossopharyngeal nerve tend not to cause major deficits, as there is bilateral corticobulbar input. A bilateral lesion will result in a pseudobulbar palsy. This is described in more detail in the Vagus Nerve chapter. A variety of causes of glossopharyngeal nerve palsy can be from intracranial tumors (cerebellopontine angle) or neck tumors.

1. *Glossopharyngeal neuralgia*
   This is a rare condition, with an unknown cause, although may be related to compression of the glossopharyngeal nerve by a nearby blood vessel, base of skull tumor around the course of the nerve, or a tumor within the throat compressing on its extracranial part. It typically presents with pain at the back of the throat, tongue, and ear. These episodes are triggered by a variety of activities involving using the mouth, for example, eating/chewing, swallowing, laughing, and speech. The episodes tend to get worse as time progresses. If it involves compression of the vagus nerve too, it can present with pain and bradycardia and in extreme cases asystole. Syncope may also be encountered with vagus nerve involvement. If there is a treatable cause, that should be managed first, as well as providing analgesia and perhaps, as with trigeminal neuralgia, an antiepileptic agent like carbamazepine.

2. *Jugular foramen syndrome*

Jugular foramen syndrome, also known as *Vernet's syndrome*, is a condition which affects the glossopharyngeal, vagus, and accessory nerves as they enter the jugular foramen. Compression of these nerves, and perhaps also the hypoglossal nerve, is caused by a variety of neoplasms, trauma, and infections. It leads to a wide variety of signs and symptoms. Typically this will involve loss of taste to the posterior one-third of the tongue, dysphagia, vocal paralysis, and anesthesia of the larynx and pharynx and weakness or paralysis of the trapezius and sternocleidomastoid muscles (Greenberg, 2006). Treatment is aimed at an identifiable cause.

2.4.10 Vagus Nerve

The vagus nerve is the tenth cranial nerve. Like the glossopharyngeal nerve, again, it is a rather complex nerve having four nuclei and five different types of fibers in it. These convey information related to sensory muscular activity and autonomic functions. Its name comes from the Latin word *vagary*, meaning wandering. It has the longest course of the cranial nerves and is extensively distributed, especially below the level of the head. It contains the following types of fibers:

a. *Branchial motor*

Supplying muscles of the pharynx and larynx.

b. *Visceral sensory*

This component of the vagus nerve is responsible for transmitting information from a wide variety of anatomical sites including the heart and lungs, pharynx and larynx, and upper part of the gastrointestinal tract.

c. *Visceral motor*

The visceral motor component carries parasympathetic fibers from the smooth muscle of the upper respiratory tract, heart, and gastrointestinal tract

d. *Special sensory*

The special sensation conveyed by the vagus nerve is for taste from the palate and epiglottis.

e. *General sensory*

The general sensory component of the vagus nerve is concerned with information from parts of the ear and the dura within the posterior cranial fossa.

The vagus nerve arises from the medulla as several rootlets. It passes towards the jugular foramen found between the glossopharyngeal and spinal accessory nerves. Like the glossopharyngeal nerve, two ganglia related to the vagus nerve are found here—the superior and inferior ganglia. The vagus nerve then descends in the carotid sheath sandwiched between the internal jugular vein and the internal and external carotid arteries. As it further descends, it is related to the internal jugular vein and the common carotid artery. Then, the right and left vagus nerves have very different anatomical pathways.

On the right side, the vagus nerve passes anterior to the right subclavian artery and posterior to the superior vena cava. At the point where it is closely related to the subclavian artery, it gives off its recurrent laryngeal branch. This branch passes under the artery then posterior to it. It then ascends between the trachea and esophagus, both of which it supplies at that point. The right recurrent laryngeal nerve then passes closely related to the inferior thyroid artery. It enters the larynx behind the cricothyroid joint and deep to the inferior constrictor. The recurrent laryngeal nerve conveys sensory information from below the level of the vocal folds, and all of the muscles of the larynx on that side, except cricothyroid.

The left vagus nerve descends towards the thorax passing between the common carotid and subclavian arteries, passing posterior to the brachiocephalic vein. It gives off branches here to the esophagus, lungs, and heart. It then passes to the left side of the arch of the aorta. From here, the recurrent laryngeal nerve is given off which descends underneath the arch of the aorta to ascend in the groove between the esophagus and trachea. As it does so, it gives off branches to the aorta, heart, esophagus, and trachea.

Meningeal Branch
This branch arises at the superior ganglion but contains the first and second cervical spinal nerves supplying the dura in the posterior cranial fossa.

Auricular Branch
This branch also arises from the superior ganglion and is joined by a branch from the glossopharyngeal nerve. It also may have a communication with a branch of the facial nerve, and supplies the auricle (cranial surface) and the tympanic membrane as well as the floor of the external auditory meatus.

Pharyngeal Branches

These branches, of which there can be several, help form the pharyngeal plexus (which also contains the glossopharyngeal nerve and the sympathetic branches) on the constrictor muscles. They supply all the muscles of the soft palate (apart from tensor veli palatini which is supplied by the medial pterygoid nerve (from the mandibular division of the trigeminal nerve)) and pharynx (apart from stylopharyngeus, which is supplied by the glossopharyngeal nerve). Most fibers are derived from the internal branch of the accessory nerve.

Superior Laryngeal Nerve

This branch descends closely related to the pharynx and has two branches—an internal and external component. The internal laryngeal nerve supplies sensation to the mucosa from the epiglottis to just *above* the level of the vocal folds. (The recurrent laryngeal nerve supplies sensation from the rest of the larynx *below* the level of the vocal folds). It pierces the thyrohyoid membrane above the superior laryngeal artery. The other branch of the superior laryngeal nerve, the external laryngeal nerve, passes under sternothyroid deep to the superior thyroid artery. It supplies the cricothyroid and the inferior constrictor muscles.

Recurrent Laryngeal Nerve

The right recurrent laryngeal nerve arises from in front of the subclavian artery. It then ascends alongside the trachea posterior to the common carotid artery. At the inferior pole of the thyroid gland, the recurrent laryngeal nerve is closely related to the inferior thyroid artery. Like the left recurrent laryngeal nerve, both of these branches have highly variable relations to the inferior thyroid artery, and the surgeon must be cautious of this when operating in and around here, for example, thyroidectomy (Yalcxin, 2006).

On the left hand side, the recurrent laryngeal nerve arises at the arch of the aorta and then passes underneath it, closely related to the ligamentum arteriosum. It then ascends in the trachea−esophageal groove.

Both recurrent laryngeal nerves pass deep to the inferior constrictor muscle and enter the larynx at the junction between the inferior cornu of the thyroid with the cricoid cartilage. They carry several types of fibers in them—motor to all the muscles of the larynx (apart from cricothyroid (supplied by the external laryngeal nerve), sensory fibers (to below the level of the vocal folds), and stretch receptors from the larynx. As it passes close to the subclavian artery or aorta, it gives off cardiac branches.

Carotid Branches
These branches are variable in number and can arise from either the glossopharyngeal or superior laryngeal nerve or the inferior ganglion.

Esophageal Branches
These branches form the esophageal plexus providing innervation to the esophagus and the posterior aspect of the pericardium.

Pulmonary Branches
The pulmonary branches exist as anterior and posterior branches, with the posterior fibers being more numerous. They ramify with the sympathetic branches supplying the bronchi and related pulmonary tissue.

Gastric Branches
The left vagus nerve supplies primarily the antero-superior aspect of the stomach. The right vagus nerve supplies the more postero-inferior region of the stomach.

Coeliac Branches
These branches go on to form the coeliac plexus, contributing also to the hepatic plexus supplying the liver. These fibers amalgamate with the fibers of the sympathetic trunk.

Renal Branches
The renal branches from the vagus nerve help to contribute to the renal plexus, which includes the splanchnic nerves. These branches go on to supply the blood vessels, glomeruli, and tubules.

Clinical Applications
As with all clinical examinations, introduce yourself to the patient, stating who you are and your purpose for meeting with the patient, especially if you are a student. Always take a comprehensive history before examining the patient. This will guide you to the most appropriate examinations that need to be conducted. Ensure you explain everything to the patient prior to doing it. This ensures a better level of trust, and excellent communication exists during the consultation.

Testing at the Bedside of the Vagus Nerve

Testing of the vagus nerve is done in exactly the same way that the glossopharyngeal nerve is tested.

1. When assessing the vagus nerve, as with the glossopharyngeal nerve, the first thing to do is simply *listening to the patient talking.* Any abnormality of the voice, for example, hoarse, whispering, or a nasal voice may give a clue as to an abnormality. Also, ask the patient if they have any difficulty in swallowing.
2. To assess the function of the vagus nerve (and the glossopharyngeal nerve) ask the patient to say "ahhhh" (without protruding their tongue) for as long as they can. Normally, the palate should rise equally in the midline. The palate (uvula) will move *towards* the side of the lesion if there is a problem with the vagus (and glossopharyngeal) nerve.
3. The gag reflex can also be assessed if relevant. You *must* tell the patient what you will do before doing this test, as it is unpleasant.
4. Using a swab, *gently* touch each palatal arch in turn, waiting each time for the patient to gag.

Tip!

Vagus nerve pathology could present with the following, affecting one or all of its branches:

1. Palatal paralysis (absent gag reflex).
2. Pharyngeal/laryngeal paralysis.
3. Abnormalities with the autonomic innervation of the organs it supplies (ie, heart, stomach (gastric acid secretion/emptying), and gut motility.

   Glossopharyngeal nerve pathology on the other hand will affect the following:

1. Dysphagia.
2. Impaired taste and sensation on the posterior one-third of the tongue.
3. Absent gag reflex.
4. Abnormal secretions of the parotid gland, though difficult to assess from the patient accurately.

Advanced Testing of the Vagus Nerve Clinically

To further assess the vagus nerve, one area that can be easily assessed more formally is by using laryngoscopy. This allows for full visualization of the vocal cords and for biopsies to be taken. If swallowing is a problem for the patient, it may be relevant, dependent on the history and examination, to undertake videofluoroscopy swallow test. Cardiovascular and gastrointestinal assessment may be considered if suggested in the clinical history and presentation.

Pathologies

An isolated nerve lesion of the vagus nerve is rare. However, if the vagus nerve is affected by pathology, whether it is neurological or trauma related, it could have widespread consequences due to the wide variety of structures it supplies.

1. *Pseudobulbar palsy*

   Pseudobulbar palsy is caused by a wide variety of conditions (neurological and vascular) but typically results from bilateral degeneration involving cranial nerve nuclei and the corticobulbar tract (pathway connecting the brainstem with the cerebral cortex). It results in the patient having difficulty swallowing (dysphagia) and difficulty in the motor aspect of speech production.

2. *Bilateral vagus nerve nucleus pathologies*

   A bilateral pathology affecting the vagus nerve will result in paralysis of the pharynx and larynx.

3. *Injury to the recurrent laryngeal nerve*

   The left recurrent laryngeal nerve runs a slightly longer course and tends to be affected more than the right for pathologies. An aneurysm of the aorta can result in compression of the left recurrent laryngeal nerve. In addition, any neck operation will place the recurrent laryngeal nerve at risk, especially if the operative field is close to the trachea-esophageal groove, heart, lungs, or esophagus. In addition, thyroidectomy can put the recurrent laryngeal nerve at risk from damage but this occurs in approximately 1% of individuals. If one recurrent laryngeal nerve is damaged, it will result in dysphonia (difficulty with speech) and hoarseness. If there is bilateral recurrent laryngeal nerve damage, it can present as a surgical emergency with inspiratory stridor, aphonia, and laryngeal obstruction. It may need to be treated by a tracheostomy in the first instance.

4. *Injury to the superior laryngeal nerve*

   Injury to the superior laryngeal nerve can occur as a complication of a thyroidectomy. It will result in paralysis of the cricothyroid muscle and anesthesia of the region above the level of the vocal folds. It tends to be, however, the external laryngeal branch that is affected. Therefore, it would affect only the cricothyroid muscle. Some patients may not have any significant consequences of this, while others may have difficulty in changing the pitch of their voice or reduced stamina in their voice. This can have disastrous consequences for those who use their voice in their careers, for example, singers and public speakers.

5. *Vagus nerve stimulation*

Some patients with epilepsy are not able to be seizure free with antiepileptic drugs alone. After trying a variety of different drugs and dosage regimens, it may be necessary to consider vagus nerve stimulation therapy. This entails inserting a small generator, similar to a pace maker which is surgically inserted to stimulate the vagus nerve 24 h a day.

6. *Highly selective vagotomy*

Vagotomy involves cutting of the vagus nerve or parts of it. Previously, it was the gold standard in severe gastric and duodenal ulcer disease. However, with extremely effective triple and quadruple therapy against Helicobacter Pylori, H2 receptor antagonists and proton pump inhibitors, the need for this surgery is not as common. Three types of vagotomy can be undertaken—truncal, selective, and highly selective vagotomy. For a highly selective vagotomy, this only involves denervation of that part of the vagus nerve which supplies the body and fundus of the stomach.

2.4.11 Accessory Nerve

The spinal accessory nerve is the 11th cranial nerve. It is a motor nerve (somatic motor) innervating two muscles—the sternocleidomastoid and trapezius. It has two components—a spinal part and a cranial part. The cranial part of the accessory nerve is from the vagus nerve. However, more recently, it has been shown that not all individuals may have a cranial root (Tubbs et al., 2014). However, when present (in the majority of cases), it joins with the spinal part of the accessory nerve for a short distance. The spinal part of the accessory nerve arises from the first five or six cervical spinal nerves. It has been shown that there is an elongated nucleus which extends from the first seven cervical vertebral levels, which provides the spinal portion of the accessory nerve (Pearson, 1937; Pearson et al., 1964). These branches arise from the lateral side of the spinal cord then form a nerve trunk. This spinal portion then ascends through the foramen magnum passing laterally to join with the cranial root.

As the two nerves join, they then pass through the jugular foramen briefly, along with the glossopharyngeal and vagus nerves. The cranial part then passes to the superior ganglion of the vagus. It then is distributed primarily in the branches of the vagus nerve, specifically the pharyngeal and recurrent laryngeal nerves.

The spinal portion then goes on to supply the sternocleidomastoid and trapezius in the neck.

The accessory nerve has two roots—a cranial and spinal division. The cranial root arises from the inferior end of the nucleus ambiguus and perhaps also from the dorsal nucleus of the vagus nucleus. The fibers of the nucleus ambiguus are connected bilaterally with the corticobulbar tract (motor neurons of the cranial nerves connecting the cerebral cortex with the brainstem). The cranial part leaves the medulla oblongata as four or five rootlets uniting together and then to join with the spinal part of the accessory nerve just as it enters the jugular foramen. At that point, it can send occasional fibers to the spinal part. It is only united with the spinal part of the accessory nerve for a brief time before uniting with the inferior ganglion of the vagus nerve. These cranial fibers will then pass to the recurrent laryngeal and pharyngeal branches of the vagus nerve, ultimately destined for the muscles of the soft palate (not tensor veli palatini (supplied by the medial pterygoid nerve of the mandibular nerve)).

The spinal root arises from the spinal nucleus found in the ventral grey column extending down to the fifth cervical vertebral level. These fibers then emerge from the spinal cord arising from between the ventral and dorsal roots. It then ascends between the dorsal roots of the spinal nerves entering the cranial cavity through the foramen magnum posterior to the vertebral arteries. It then passes to the jugular foramen, where it may receive some fibers from the cranial root. As it then exits the jugular foramen, it is closely related to the internal jugular vein. It then courses inferiorly passing medial to the styloid process and attached stylohyoid. It also is found medial to the posterior belly of digastric. The spinal root then supplies the sternocleidomastoid muscle on its medial aspect.

The cranial root then enters the posterior triangle on the neck lying on the surface of the levator scapulae at approximately midway down the sternocleidomastoid. As it passes inferiorly through the posterior triangle of the neck and just above the clavicle, it then enters the trapezius muscle on its deep surface at its anterior border. The third and fourth cervical vertebral spinal nerves also supply the trapezius forming a plexus of nerves on its deeper surface. Figs. 2.15 and 2.16 demonstrate the position of the accessory nerve and related structures.

Clinical Applications
Testing at the Bedside of the Accessory Nerve
From the clinical perspective, the accessory nerve supplies the *sternocleidomastoid* and *trapezius* muscles, and as such, it is those that are tested when assessing the integrity of the nerve.

The sternocleidomastoid muscle has two functions dependent on whether it is acting on its own or with the opposite side. If the sternocleidomastoid is acting on its own, it tilts the head to that side it contracts and, due to its attachments and orientation, rotates the head so

Figure 2.15 An unlabeled image to demonstrate the position of the accessory nerve in relation to other structures.

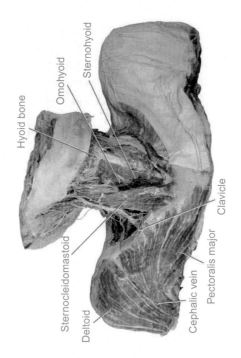

Figure 2.16 A labeled image to demonstrate the position of the accessory nerve in relation to other structures.

that the face moves in the direction of the opposite side. Therefore, if the left sternocleidomastoid muscle contracts, the face turns to the right hand side, and vice versa.

If, however, both sternocleidomastoid muscles contract, the neck flexes and the sternum is raised, as in forced inspiration.

The trapezius is an extremely large superficial muscle of the back. It comprised three united parts—superior, middle, and inferior. It is involved in two main functions dependent on if the scapula or the spine is stable. If the spinal part is stable, it helps move the scapula, and if the scapula is stable, it helps move the spine. Trapezius is involved in a variety of movements. The upper fibers raise the scapula, the middle fibers pull the scapula medially, and the lower fibers move the medial side of the scapula down. Therefore, trapezius is involved in both elevation and depression of the scapula, dependent on which part is contracting. As well as this, the trapezius also rotates and retracts the scapula.

As with all clinical examinations, introduce yourself to the patient, stating who you are and your purpose for meeting with the patient, especially if you are a student. Always take a comprehensive history before examining the patient. This will guide you to the most appropriate examinations that need to be conducted. Ensure you explain everything to the patient prior to doing it. This ensures a better level of trust, and excellent communication exists during the consultation.

Testing of the accessory nerve is done as follows:

1. *Always* inform the patient of what you will be doing, after introducing yourself and taking a detailed clinical history.
2. When examining a patient, ensure you just observe the patient and try to identify if there is any obvious deformity, or asymmetry of the shoulder and neck region. It may be that you will see an obvious weakness or asymmetrical position of the patient's neck and/ or upper limbs.
3. First, you can assess the sternocleidomastoid.
4. You can ask the patient to rotate their head to look to the left and right hand sides to identify any obvious abnormality.
5. Then, ask the patient to look to one side and test the muscle against resistance.
6. For example, if the patient looks to the right side, place the ball of your hand on their left mandible.
7. Ask the patient to press into your hand.
8. Repeat this on the opposite side.
9. Then, you need to assess the trapezius.
10. First you can ask the patient to raise their shoulders, as in shrugging.
11. Observe any gross abnormality.
12. Then, while the patient is raising their shoulders, gently press down on them as they lift their shoulders.
13. Assess any weakness which may be present, noting which side is affected.

Tip!

When assessing the function of the sternocleidomastoid and trapezius, it may help examining the unaffected side first, especially if the patient complains of pain or discomfort on one side. This helps build up trust with the patient but also minimizes causing them any pain or discomfort.

Tip!

Weakness in rotating the head to the left hand side, when examining sternocleidomastoid, suggests a pathology with the right accessory nerve, and vice versa.

Advanced Testing of the Accessory Nerve

The bedside testing should be sufficient in examining the accessory nerve, but the results of this may help direct towards any specialist investigations that may be relevant for the patient. Electromyography studies may be deemed relevant, as well as further cranial and neck investigation by means of CT and/or MRI scanning, dependent on the clinical history and physical examination findings.

Pathologies

Pathology of the accessory nerve can be divided into various points the nerve leaves the brainstem to its final destination, that is, the sternocleidomastoid and trapezius.

a. *Supranuclear lesion*

A supranuclear lesion will result in weakness of both sternocleidomastoid and trapezius muscles, due to bilateral innervation. Within the spinal cord the nuclei can be affected by polio, intraspinal tumors, or amyotrophic lateral sclerosis (motor neuron disease).

If the blood supply to the lateral part of the medulla is compromised, for example, due to occlusion of the posterior inferior cerebellar artery, a number of clinical presentations will result. Occlusion of this vessel will affect the trigeminal, glossopharyngeal, vagus, and accessory nerves. Therefore, speech and gag reflex, balance, and facial sensation will also be affected. This is classified as *Wallenberg's syndrome.*

b. *Compression at the level of the jugular foramen*

Any lesion that exists at the level of the jugular foramen can result in compression of the accessory nerve, and also the other cranial nerves entering at that point, that is, the glossopharyngeal and vagus nerves. This can range from neoplasms to vascular problems, for example, aneuryms. If the glossopharyngeal, vagus, and accessory nerves are affected together, it is referred to as the jugular foramen syndrome, or Vernet's syndrome. Typically, with Vernet's syndrome, there will be loss of taste to the posterior of the tongue, weakness, or paralysis of the sternocleidomastoid and trapezius muscles and dysphagia, vocal paralysis, and anesthesia of the larynx and pharynx.

c. *Lesion of the nerve in the posterior triangle.*
An isolated accessory nerve lesion is not common. Typically, it is damaged in surgical procedures involving its passageway through the posterior triangle of the neck, for example, neck dissection or surgical biopsy of tissue.

Torticollis

This condition is a dystonia presenting with an asymmetrical head/neck with the head tilted to one side. A variety of causes exist for it, for example, congenital, trauma-related, infections of the pharynx, base of skull tumors, and cervical vertebral abnormalities. Treatment tends to be alleviating any related pain and gentle exercises to release the stiffness in the neck.

2.4.12 Hypoglossal Nerve

The hypoglossal nerve is the 12th cranial nerve. It supplies all but one of the intrinsic and extrinsic muscles of the tongue and is a general somatic efferent nerve (somatic motor). It exits the skull at the hypoglossal canal and gives off the meningeal, thyrohyoid, and lingual branches, as well as a component of the ansa cervicalis.

The hypoglossal nerve arises from the hypoglossal nucleus. This is found at the full length of the medulla, close to the midline. The roots of the hypoglossal nerve itself arise between the olive and the pyramid. These roots then unite to form two bundles which then pass through the hypoglossal canal (in the occipital bone) by piercing the dura mater. It is at this point the two bundles unite to form the full hypoglossal nerve as a single entity.

After leaving the hypoglossal canal, the hypoglossal nerve receives twigs from C1 and C2 containing general somatic motor fibers, as well as some general sensory fibers from the C2 ganglion. It is these fibers that go on further in the neck to supply the strap muscles.

The hypoglossal nerve then continues its journey to the tongue by passing posterior to the internal carotid artery and the glossopharyngeal and vagus nerves. It then passes inferiorly sandwiched between the internal jugular vein and the internal carotid artery. It then loops forward over the occipital artery receiving some fibers from the pharyngeal plexus at that point. As it descends further, it passes over several arteries—the external and internal carotid, and lingual arteries. It lies on the hyoglossus inferior to the lingual nerve and submandibular duct then passes inferior to the mylohyoid and digastric. From the hypoglossal nerve are several important branches, which will be discussed later—the meningeal, thyrohyoid, and muscular branches as well as a component to the superior root of the ansa cervicalis. Figs. 2.17 and 2.18 demonstrate the hypoglossal nerve and related structures.

Meningeal Branches
These fibers pass upwards through the hypoglossal canal to supply the dura mater within the posterior cranial fossa.

Thyrohyoid Branch
This branch arises close to the hyoglossus, passing across the greater cornu of the hyoid bone and supplies the thyrohyoid.

Figure 2.17 An unlabeled image to demonstrate the hypoglossal nerve in relation to other structures.

Muscular (lingual) Branches

These branches supply the intrinsic muscles of the tongue but also a number of other muscles including styloglossus, genioglossus, geniohyoid, and hyoglossus. However, the fibers to the geniohyoid hitchhike with the hypoglossal nerve but originate from the first cervical nerve.

Superior Root of Ansa Cervicalis

The ansa cervicalis is a loop of nerves which are formed for the first three cervical nerves (C1−C3) and innervate the infrahyoid ("strap") muscles. The ansa cervicalis has a superior and inferior root.

The superior root of the ansa cervicalis is created by the first cervical nerve (C1). This root passes in the fibers of the hypoglossal nerve before coming away from it within the carotid triangle and thus forms its superior root. It passes round the occipital artery and passes down to the carotid sheath. It innervates the omohyoid muscle's superior belly as well as the superior portion of the sternothyroid and the sternohyoid muscles and unites with the inferior root of the ansa cervicalis.

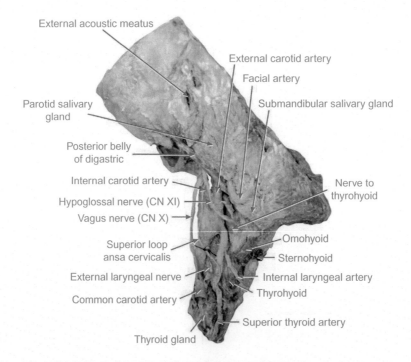

Figure 2.18 A labeled image to demonstrate the hypoglossal nerve in relation to other structures.

The inferior root of the ansa cervicalis is formed by the unification of the second and third cervical nerves. It courses inferiorly on the outer aspect of the internal jugular vein and then crosses roughly midway down this vessel to pass anterior to this vein. It then passes an anterior direction to unite with the superior root of the ansa cervicalis anterior to the common carotid artery. The inferior root of the ansa cervicalis may pass between the common carotid artery and the internal jugular vein. All the infrahyoid ("strap") muscles are innervated by the ansa cervicalis, except the thyrohyoid which is innervated by the first cervical nerve which accompanies the hypoglossal nerve for a short distance.

Clinical Applications
As with all clinical examinations, introduce yourself to the patient, stating who you are and your purpose for meeting with the patient, especially if you are a student. Always take a comprehensive history before examining the patient. This will guide you to the most appropriate examinations that need to be conducted. Ensure you explain everything to the patient prior to doing it. This ensures a better level of trust, and excellent communication exists during the consultation.

Testing at the Bedside of the Hypoglossal Nerve
There are two simple methods which can be used to test the integrity of the hypoglossal nerve, and the tongue it supplies.

1. The first thing to do is simply observe the tongue at rest in the oral cavity. Ask the patient to open their mouth. Observe for any muscle wasting or fasciculations which may be present, as well as any surface lesions. If there is pathology, the tip of the tongue tends to rest to the normal side due to the unopposed pull of the muscle on that side.
2. Then, ask the patient to stick out their tongue. This will assess the muscle function on both sides. If there is a pathology present, the tongue will appear weakened on the affected side and *the tongue will point to the side of pathology*. This is due to the unopposed muscle power of the normal side.

Advanced Testing of the Hypoglossal Nerve
Further assessment of the hypoglossal nerve will be directed by the initial presentation, the history taken from the patient and a clinical examination. From this, it will direct the examining doctor to the

relevant investigations which may include, but not limited to, CT, MRI, x-ray, hematological tests, or lumbar puncture. Although idiopathic lesions of the hypoglossal nerve can occur, they are very rare, and in most cases, indicate a space occupying lesion, vascular, neurological, or trauma-related cause (Freedman et al., 2008).

Pathologies
Pathologies of the hypoglossal nerve can be subdivided based on the level of the pathology.

1. *Supranuclear pathologies*
   These types of lesions tend to be highly variable in their presentation, but generally, they produce a very weak and transient weakness of the tongue musculature.
2. *Hypoglossal nucleus pathology*
   These lesions tend to be bilateral as the hypoglossal nucleus is located close to the midline, and therefore in close proximity to the one of the opposite side. A variety of presentations can result from a hypoglossal nucleus lesion—weakness and atrophy, complete paralysis or fasciculations (twitching).
   Hypoglossal nucleus pathologies can arise for a variety of causes, for example, due to spinal tumors, infarction, or neurological conditions like syringobulbia, amyotrophic lateral sclerosis, or polio.
3. *Hypoglossal nerve pathology*
   The hypoglossal nerve can be affected anywhere along its length. Tumors tend to be the most common pathology which affects the hypoglossal nerve (Keane, 1996). These can occur anywhere along it course, but specifically can happen at the jugular foramen. At that site, the accessory and glossopharyngeal nerves could be affected. In addition to this trauma, infection (eg, basilar meningitis) or neurological conditions (multiple sclerosis, Guillain-Barré neuropathy) are possible causes which will affect the hypoglossal nerve in its passage to the muscles of the tongue.

2.5 CUTANEOUS NERVES

Cutaneous innervation of the face overlaps that of the neck. Indeed, sensory innervation of the face arises from the terminal branches of the three main divisions of the trigeminal nerve, that is, the ophthalmic, maxillary, and mandibular divisions of this nerve.

The trigeminal nerve arises from the lateral aspect of the pons comprised a large sensory root and a smaller motor root. Cell bodies of the trigeminal nerve are located in the trigeminal ganglion with a lesser amount in the mesencephalic trigeminal nucleus.

It is the peripheral processes of the ganglion that forms the ophthalmic and maxillary nerves and the sensory part of the mandibular nerve. In addition, within the mandibular nerve, proprioceptive fibers are present from the mesencephalic nucleus.

Central processes of the trigeminal ganglion enter the pons and then pass to the spinal and pontine trigeminal nuclei. It is the large fibers for discriminative touch that terminate in the pontine trigeminal nucleus. Indeed, the pontine trigeminal nucleus is referred to as the chief or principal sensory nucleus.

However, a smaller number of fibers pass caudally towards the spinal cord to the spinal trigeminal nucleus. The spinal trigeminal nucleus is responsible for conveying information related to light touch, pain, and temperature.

The information conveyed in the spinal trigeminal tract also includes input from the outer aspect of the ear, posterior one-third of the tongue (mucosa of), pharynx, and larynx. This means there is an input related to sensation also from these sites and is related to the facial, glossopharyngeal, and vagus nerves, respectively. In the sensory root and the spinal cord portion of the trigeminal nerve, there is a spatial arrangement of the fibers. The mandibular fibers are initially ventral and the ophthalmic fibers dorsal, with the maxillary fibers lying between these. On approach to the brainstem, there is a rotation of the fibers to lie the opposite way, that is, the mandibular fibers end up dorsal and the ophthalmic fibers ventral, and again the maxillary fibers sandwiched between the two.

In addition to this, the mesencephalic trigeminal nucleus extends from the pontine trigeminal nucleus to the midbrain. This has two processes—a central part and a peripheral portion. The peripheral branches of the mesencephalic trigeminal nucleus pass within the mandibular nerve and terminate in proprioceptive receptors beside the teeth of the mandible and in the muscles of mastication (ie, temporalis, masseter, and the pterygoid muscles). Occasional fibers also pass to the maxillary division ending in the hard palate adjacent to the teeth of the maxilla.

The central branches of the mesencephalic trigeminal nucleus terminate in either the motor nucleus of the trigeminal nerve or the reticular formation (and then onwards to the thalamus).

The bulk of the motor root of the trigeminal nerve contain fibers from the trigeminal motor nucleus. This nucleus, found medial to the chief or principal sensory nucleus (ie, pontine trigeminal nucleus), supplies the muscles of mastication (temporalis, masseter, and the pterygoids (lateral and medial)). It also supplies the anterior belly of digastric, mylohyoid, tensor veli tympani, and tensor tympani. The trigeminal motor nucleus receives afferent information from the corticobulbar tract (that white matter pathway connecting the cerebral cortex to the brainstem). Afferent information also arrives from the sensory trigeminal nuclei. This pathway deals with the stretch reflex and the jaw-opening reflex.

On leaving the brainstem, the motor fibers of the trigeminal nerve pass below the ganglion along the floor of Meckel's cave (named after the German anatomist Johann Friedrich Meckel, the elder (to avoid confusion with his famous grandson, also anatomists). These fibers are only present in the mandibular division of the trigeminal nerve and become related to the sensory fibers as the whole nerve passes through the foramen ovale. It goes on to supply the muscles of mastication, and some of the smaller muscles previously described. Now, the terminal branches of the trigeminal nerve innervating the face will be dealt with.

2.5.1 Supraorbital Nerve

The supraorbital nerve is the direct continuation of the frontal nerve. It leaves the orbit through the supraorbital notch, or the supraorbital foramen. Knize (1995) had shown that from the orbital rim, there were two constant divisions of this nerve. The first was a medial or superficial division, which went over frontalis, providing sensory innervation to the skin of the forehead and to the front margin of the scalp. The second division was found deeper and lateral and ran across the outer aspect of the forehead between the epicranial aponeurosis and the pericranium supplying sensory innervation to the frontoparietal region of the scalp.

2.5.2 Supratrochlear Nerve

The supratrochlear nerve passes anteriorly and medially passing superior to the trochlea of the superior oblique. It gives off a small filament which descends to join with the infratrochlear branch of the nasociliary nerve. The supratrochlear nerve then leaves the orbit from between the supraorbital foramen and the trochlea, passing in a superior direction over the forehead. Closely related to the supratrochlear nerve is the supratrochlear branch of the ophthalmic artery. Small fibers of the nerve pass to the skin of the upper aspect of the eyelid and then it passes under the cover of the frontal belly of the occipitofrontalis muscle and the corrugator muscles. It innervates the skin of the inferior part of the forehead close to the median plane.

2.5.3 Lacrimal Nerve

The lacrimal nerve is the smallest of the main branches that arise from the ophthalmic nerve. Occasionally it will receive a fiber from the trochlear nerve. The lacrimal nerve then passes into the orbit through the superior orbital fissure, specifically its lateral part. It then courses along the superior border of the lateral rectus muscle, alongside the lacrimal artery. The lacrimal nerve will then enter the lacrimal gland, after it receives a small branch from the zygomaticotemporal branch of the maxillary nerve. Inside the lacrimal gland, the lacrimal nerve goes off a few small fibers to the gland itself but also to the skin of the upper eyelid. It then passes into the orbital septum and terminates in the skin of the upper eyelid, where it will ramify with branches of the facial nerve.

2.5.4 Infratrochlear Nerve

The infratrochlear nerve arises from the nasociliary nerve close to the anterior ethmoidal foramen. It courses anteriorly along the inner wall of the orbit, above the superior border of the medial rectus muscle. Close to this point, the infratrochlear nerve then receives a small branch from the supratrochlear nerve. The infratrochlear nerve then goes on to innervate the skin of the eyelids as well as the side of the nose superior to the medial angle of the eye. It also innervates the lacrimal caruncle and sac as well as the conjunctiva.

2.5.5 External Nasal Nerve

The nasociliary nerve is the sensory nerve to the eye. It enters the orbit through the superior orbital fissure, inside the cone formed by the muscles of the globe. It is on a lower plane, therefore, than the lacrimal and frontal nerves. It lies between the two divisions of the oculomotor nerve. It passes anteriorly inferior to the superior rectus and crosses the optic nerve with the ophthalmic artery. At the medial aspect of the orbit, it lies between the superior oblique and the medial rectus. It is continued as the anterior ethmoidal nerve.

The nasociliary nerve gives off several smaller branches:

a. *Communicating branch*
   This communicating branch passes to the ciliary ganglion (see "Abducent Nerve").
b. *Long ciliary nerves*
   These can exist either singly or two nerves and convey sympathetic fibers to the dilator pupillae and afferent fibers from the uvea and cornea.
c. *Infratrochlear nerve*
   The infratrochlear nerve passes to the eyelids, skin of the nose, and the lacrimal sac.
d. *Posterior ethmoidal nerve*
   The posterior ethmoidal nerve is frequently absent. When present, it provides sensory innervation to the posterior ethmoidal and sphenoidal sinuses.
e. *Anterior ethmoidal nerve*
   The anterior ethmoidal nerve is seen as the continuation of the nasociliary nerve. The anterior ethmoidal nerve passes through the anterior ethmoidal foramen entering into the anterior cranial fossa. It passes into the nasal cavity dividing into internal nasal branches. These branches supply the walls of the nasal cavity. One of the branches passes to the skin of the nose as an external nasal branch.

In its course, the nasociliary nerve, together with its continuation, the anterior ethmoidal nerve, traverses in succession the middle cranial fossa, the orbit, the anterior cranial fossa, nasal cavity, and the external aspect of the nose.

The termination of the anterior ethmoidal nerve is by the external nasal branch which arises at the inferior aspect of the nasal bone.

It courses under the transverse portion of the nasalis muscle and innervates the skin of the ala of the nose as well as the apex and the vestibule of the nose.

Variation does exist in the distribution of the external nasal nerve. The external nasal nerve has been found to either continue as a single long branch, or divide into two branches at it approaches the apex of the nose.

Clinical Applications
When undertaking any clinical history taking or examination, you should always do the following, and follow a logical and systematic format:

1. Introduce yourself to the patient.
2. Advise them of what position you hold, for example, student, specialty grade, consultant, etc.
3. Your reason for consulting with them or to find out why they have presented to you.
4. Always take a thorough and detailed history, which will be guided by the presenting signs and symptoms.
5. When examining the patient, always tell them what you will ask them to do, or what region of the body you will be examining, with specific instructions and ensure they give consent.

Surgical Damage to the External Nasal Nerve
It has been shown that careless endonasal dissection or rhinoplasty surgical procedures can place the external nasal nerve at risk from damage. It has been shown that to avoid injuring the external nasal nerve for any procedures along its course, deep incisions into the cartilage or intercartilaginous incisions where the dissection was restricted to within 6.5 mm from the midline, and also limiting dorsal nasal grafts to 13 mm at the rhinion (Han et al., 2004).

2.5.6 Zygomaticotemporal Nerve
The zygomatic nerve originates in the pterygopalatine fossa, and it then passes into the orbit via the inferior orbital fissure. It then runs on the outer wall of the orbit and terminates as two branches—the zygomaticotemporal and the zygomaticofacial nerves.

The zygomaticotemporal nerve runs along the lower outer aspect of the orbit and provides a branch to the lacrimal nerve. It then passes through a small canal in the zygomatic bone and then arrives into the temporal fossa. It passes superiorly between the bone and the temporalis muscle. It will then go through the temporal fascia just a couple of centimeters superior to the zygomatic arch and innervates the skin of the temple region. It anastomoses with the facial nerve and also with the auriculotemporal nerve form the mandibular division of the trigeminal nerve. When it pierces the temporal fascia, it also sends a small branch between the two layers of temporal fascia to reach the outer aspect of the eye.

2.5.7 Zygomaticofacial Nerve
The zygomaticofacial nerve runs along the lower outer aspect of the orbit, and arrives onto the surface of the face through a foramen in the zygomatic bone. It then passes through the orbicularis oculi and innercates the skin over the prominence of the cheek. It ramifies with the zygomatic branch of the facial nerve and also the palpebral branches from the maxillary nerve of the trigeminal nerve.

2.5.8 Infraorbital Nerve
The maxillary division of the trigeminal nerve, also referred to as the maxillary nerve, is a purely sensory (afferent) nerve. It is the medium-sized branch of the trigeminal nerve between the smaller ophthalmic division and the largest mandibular division. After emerging from the trigeminal ganglion, it passes to the pterygopalatine fossa, passing to the posterior surface of the maxilla before passing through the foramen rotundum and entering the orbit through the inferior orbital fissure, running here to terminate on the anterior aspect of the skull at the infraorbital foramen. The maxillary nerve also passes through the cavernous sinus. In general, it supplies the teeth of the maxilla, skin from the lower eyelid above to the superior aspect of the mouth below, as well as the nasal cavity and the paranasal sinuses. The specific branches of the maxillary nerve are as follows:

1. *Middle meningeal branch*—to supply the dura.
2. *Alveolar nerves* (anterior, middle, and posterior superior alveolar nerves)—provides sensory innervation to all of the upper teeth in the maxilla, as well as the gingiva.

3. *Zygomatic nerve*—supplying the skin of the side of the forehead (via the zygomaticotemporal nerve) and the area over the prominence of the cheek (maxilla) anteriorly (via the zygomaticofacial nerve). This nerve also carries with it parasympathetic postganglionic fibers from the facial nerve to innervate the lacrimal gland.

4. *Palatine nerves* (greater and lesser palatine nerves, as well as the nasopalatine nerve)—to supply the gingiva, mucous membranes of the roof of the mouth (via the greater palatine nerve), soft palate (including uvula), and tonsils (via the lesser palatine nerve) and the palatal structures around the superior anterior six teeth (via the nasopalatine nerve).

5. *Pharyngeal nerve*—a small branch which passes to the area behind the auditory tube supplying the nasopharynx for its mucosa.

6. *Infraorbital nerve*—this nerve exits through the infraorbital foramen of the maxilla to supply the lower eyelid and superior lip.

7. *Inferior palpebral nerve*—supplies the skin of the inferior eyelid as well as the conjunctiva.

8. *Superior labial nerve*—supplies the skin of the superior lip and the mucosa of the mouth at this point.

9. *External nasal branches*
   The infraorbital nerve is seen as the continuation of the maxillary nerve from the trigeminal nerve. It enters the orbit through the inferior orbital fissure and then occupies the infraorbital groove, canal, and foramen. It terminates on the face by dividing into several branches namely the inferior palpebral (which innervates the conjunctiva and the skin of the lower eyelid), nasal (innervating the skin of the nose), and the superior labial (which innervates the mucous membrane of the mouth and the skin of the lip). A middle superior alveolar branch typically arises from the infraorbital nerve and passes in the anterior, lateral, or the posterior wall of the maxillary sinus, and then passes to the premolar part of the superior dental plexus. An anterior superior alveolar branch arises from the infraorbital nerve in the infraorbital canal, and by means of a tortuous canal, descends along the anterior wall of the maxillary sinus. It forms the superior dental plexus and provides innervation to the canine and incisor teeth. Its terminal branches emerge close to the nasal septum and innervate the floor of the nose. The superior dental plexus is found in part on the posterior surface of the maxilla and in part in the bony canals of the lateral and anterior aspects of the maxilla. It is formed by the anterior and posterior, and when present, the middle alveolar nerves.

2.5.9 Auriculotemporal Nerve

The auriculotemporal nerve arises typically by two roots which encircle the middle meningeal artery. The nerve passes posteriorly deep to the lateral pterygoid and sandwiched between the sphenomandibular ligament and the neck of the mandible. The auriculotemporal nerve is intimately associated with the parotid gland and it then courses superiorly posterior to the TMJ. It crosses the zygoma and lies behind the superficial temporal artery. The terminal branches of the auriculotemporal nerve go on to innervate the scalp.

In addition, the auriculotemporal nerve anastomoses with the facial nerve and also the otic ganglion. The two branches which communicate with the facial nerve do so at the posterior aspect of the master. The fibers from the auriculotemporal nerve which unite with the otic ganglion do so close to their origin.

The auriculotemporal nerve has five main branches. These are the anterior auricular, articular, parotid, superficial temporal, and also branches to the external auditory meatus. The anterior auricular branches innervate the skin overlying the tragus, as well as the adjacent part of the helix. The articular branches go on to innervate the posterior aspect of the TMJ, which it is closely related to. The parotid branches provide a secretomotor innervation to the parotid gland. The preganglionic component of this originally comes from the glossopharyngeal nerve via its tympanic branch. This travels with the lesser petrosal nerve to the otic ganglion. The postganglionic fibers then pass with the auriculotemporal nerve and therefore innervate the parotid gland. In addition to this, the parotid branches also contain vasomotor fibers innervating the vasculature within the parotid gland. The superficial temporal branches run with the superficial temporal arteries. These branches innervate the skin over the temple and anastomose with the facial and also the zygomaticotemporal nerves. The branches to the external auditory meatus, which are typically two in number, run between the cartilaginous and bony ear canals and innervate the skin of the meatus, but also provide innervation to the tympanic membrane.

2.5.10 Buccal Nerve

It should be noted here as a word of caution that confusion may arise when referring to the buccal nerve. That is because there is a buccal branch of the facial nerve, but also a buccal branch of the marginal mandibular nerve.

Buccal Nerve From the Mandibular Nerve

The mandibular division of the trigeminal nerve, also referred to as the mandibular nerve, is a mixed sensory and branchial motor nerve. It is also the largest of the three branches of the trigeminal nerve. The sensory root arises from the lateral aspect of the ganglion, with the motor division lying deeper. In general, the mandibular nerve supplies the lower face for sensation over the mandible, including the attached teeth, the TMJ and the mucous membrane of the mouth as well as the anterior two-thirds of the tongue (the posterior one third is supplied by the glossopharyngeal nerve). It also supplies the muscles of mastication which are the medial and lateral pterygoids, temporalis, and masseter. It also supplies some smaller muscles namely the tensor veli tympani, tensor veli palatini, mylohyoid, and the anterior belly of digastric.

The mandibular nerve enters the infratemporal fossa and passes through the foramen ovale in the sphenoid bone, and divides at that point into a smaller anterior and a larger posterior trunk. The main trunk gives off two branches at this point. The first is a meningeal branch and passes through the foramen spinosum to receive innervation from the meninges of the middle cranial fossa. The second small branch, a muscular branch, supplies the medial pterygoid and also a twig to the otic ganglion to supply the tensor veli palatini and the tensor tympani. Two main divisions arise from the main trunk of the mandibular nerve after these two smaller branches have been given off: an anterior and posterior division.

1. *Anterior division:*
   a. *Masseteric nerve*—passing posterior to the tendon of temporalis, this branch approaches the masseteric muscle on its deep aspect.
   b. *Deep temporal nerves*—two branches arise generally from this— an anterior and posterior division. Sometimes, a third (intermediate) branch may be found.
   c. *Lateral pterygoid nerve*—this branch enters the deep surface of the muscle.
   d. *Buccal branches*

Tip!

DO NOT CONFUSE THIS WITH THE BUCCAL BRANCH OF THE FACIAL NERVE WHICH CONVEYS MOTOR INFORMATION TO BUCCINATOR.

This buccal branch from the anterior division of the mandibular nerve passes between the lateral pterygoid heads then inferior to the temporalis tendon passing to the buccal membrane to receive sensory information from that site, skin over the cheek, as well as the second and third molars. Indeed, some of the branches of the buccla nerve from the marginal mandibular nerve anastomose with the buccal branches of the facial nerve.

2. *Posterior division*
   a. *Auriculotemporal nerve*—this branch has two roots and is closely related to the middle meningeal artery. It passes postero-superiorly behind the TMJ. It is closely related to the superficial temporal vessels and gives off secretomotor fibers to the parotid gland, before reaching the temporal region and receives sensory information from here, as well as the superior half of the pinna and the external auditory meatus.
   b. *Lingual nerve*—this nerve carries two parts of information—that is related to the trigeminal nerve for sensation from the tongue (general somatic afferent), but also carries with it a branch from the facial nerve—the chorda tympani nerve for special sensory fibers of taste from the front two-thirds of the tongue, but also preganglionic parasympathetic fibers to the submandibular ganglion. The lingual nerve first passes below the lateral pterygoid muscle, receives the chorda tympani nerve, then passes between the medial pterygoid, and then passes towards the tongue.
   c. *Inferior alveolar nerve*—this is a dual motor and sensory nerve. It passes on the medial aspect of the lateral pterygoid and just before entering the mandibular foramen, it gives rise to the motor branches to the mylohyoid and anterior belly of digastric muscles. After it enters the mandibular foramen, it supplies all the lower teeth and the alveolar ridges. As it passes anteriorly, it then exits the mandible via the mental foramen and becomes the mental nerve. This nerve supplies sensation to the skin over the chin. This nerve is crucial in dental practice, as it is the nerve that is anesthetized as it enters the mandibular foramen to provide complete nerve block if procedures are to be undertaken on the lower teeth or related structures.

Buccal Branch of the Facial Nerve
The branches of the facial nerve can be divided into intratemporal and extratemporal branches. The intratemporal branches of the facial nerve are as follows:

- *Greater petrosal nerve*
  This branch arises from the geniculate ganglion and is joined by the nerve of the pterygoid canal. It contains secretomotor fibers for the lacrimal and nasal glands.
- *Nerve to stapedius*
  Supplies the stapedius muscle, which is responsible for "dampening down" loud noises protecting the middle and inner ear structures
- *Chorda tympani nerve*
  This nerve joins the lingual nerve (of the mandibular division of the trigeminal nerve—see chapter on trigeminal nerve) and is distributed to the anterior two-third of the tongue. It contains the following:
  a. Taste and sensation fibers from the front two-third of the tongue and the soft palate.
  b. Preganglionic secretory and vasodilator fibers that synapse in the submandibular ganglion.

2.5.11 Mental Nerve
The inferior alveolar nerve is a dual motor and sensory nerve. It passes on the medial aspect of the lateral pterygoid and just before entering the mandibular foramen, it gives rise to the motor branches to the mylohyoid and anterior belly of digastric muscles. After it enters the mandibular foramen, it supplies all the lower teeth and the alveolar ridges. As it passes anteriorly, it then exits the mandible via the mental foramen and becomes the mental nerve. This nerve supplies sensation to the skin over the chin. This nerve is crucial in dental practice, as it is the nerve that is anesthetized as it enters the mandibular foramen to provide complete nerve block if procedures are to be undertaken on the lower teeth, or related structures.

REFERENCES

Benzel, E.C., 2012. The Cervical Spine, fifth ed. Lippincott Williams and Wilkins, Philadelphia, USA.

Chakravarthi, S., Kesav, P., Khurana, D., 2013. Wall-eyed bilateral internuclear ophthalmoplegia with vertical gaze palsy. QJM. Available from: http://dx.doi.org/10.1093/qjmed/hct021. Available at: <http://qjmed.oxfordjournals.org/content/early/2013/01/25/qjmed.hct021.full> (Accessed January 09, 2014).

Chen, C.M., Lin, S.H., 2007. Wall-eyed bilateral internuclear ophthalmoplegia from lesions at different levels in the brainstem. J. Neuroophthalmol. 27, 9–15.

Elaimy, A.L., Hanson, P.W., Lamoreaux, W.T., Mackay, A.R., Demakas, J.J., Fairbanks, R.K., et al., 2012. Clinical outcomes of gamma knife radiosurgery in the treatment of patients with trigeminal neuralgia. Int. J. Otolaryngol. 2012, 13, Article ID 919186, http://dx.doi.org/10.1155/2012/919186.

Fisch, U.P., 1973. Excision of the Scarpa's ganglion. Arch. Otolaryngol. 97, 147–149.

Freedman, M., Jayasundara, H., Stassen, L.F.A., 2008. Idiopathic isolated unilateral hypoglossal nerve palsy: a diagnosis of exclusion. Oral Surg. Oral Med. Oral Pathol. Oral Radiol. Endod. 106, e22–e26.

Han, S.K., Shin, Y.W., Kim, W.K., 2004. Anatomy of the external nasal nerve. Plast. Reconstr. Surg. 114 (5), 1055–1059.

Keane, J.R., 1996. Twelfth-nerve palsy. Analysis of 100 causes. Arch. Neurol. 53, 561–566.

Knize, D.M., 1995. A study of the supraorbital nerve. Plast. Reconstr. Surg. 96, 564–569.

Manzoni, G.C., Torelli, P., 2005. Epidemiology of typical and atypical craniofacial neuralgias. Neurol. Sci. 26 (Suppl. 2), s65–s67.

Marklund, S., Wänman, A., 2007. Incidence and prevalence of temporomandibular joint pain and dysfunction. A one-year prospective study of university students. Acta Odontol. Scand. 65, 119–127.

Paturet, G., 1951. Traite D'anatomie Humaine, vol. 1. Masson et Cie, Paris.

Pearson, A., 1937. The spinal accessory nerve in human embryos. J. Comp. Neurol. 68, 243–266.

Pearson, A.A., Sauter, R.W., Herrin, G.R., 1964. The accessory nerve and its relation to the upper spinal nerves. Am. J. Anat. 114, 371–391.

Shoja, M.M., Oyesiku, N.M., Griessenauer, C.J., Radcliff, V., Loukas, M., Chern, J.J., et al., 2014. Anastomoses between lower cranial and upper cervical nerves. Clin. Anat. 27, 118–130.

Spoendlin, H., 1988. In: Alberti, P.W., Ruben, R.J. (Eds.), Biology of the Vestibulocochlear Nerve, first ed. Churchill Livingstone, New York, NY, pp. 117–150.

Tubbs, R.S., Benninger, B., Loukas, M., Gadol, A.A.C., 2014. Cranial roots of the accessory nerve exist in the majority of adult humans. Clin. Anat. 27, 102–107.

Walker, H.K., Hall, W.D., Hurst, J.W., 1990. Clinical Methods, The History, Physical, and Laboratory Examinations, third ed. Butterworths, Boston, ISBN-10: 0-409-90077-X. Found at: <http://www.ncbi.nlm.nih.gov/books/NBK201/> (Accessed October 5, 2015.).

Wall, M., Wray, S., 1983. The one-and-a-half syndrome: a unilateral disorder of the pontine tegmentum—a study of 20 cases and review of the literature. Neurology 33, 971–978.

Warwick, R., 1953. Representation of the extraocular muscles in the oculomotor nuclei of the monkey. J. Comp. Neurol. 98, 449–503.

Williams, Warwick, 1980. Gray's Anatomy, 36th ed. Churchill Livingstone, New York, NY.

Yalcxin, B., 2006. Anatomic configurations of the recurrent laryngeal nerve and inferior thyroid artery. Surgery 139, 181–187.

CHAPTER 3

Neck

3.1 INTRODUCTION

The neck is a transitional area between the thorax and the clavicles below and the base of the cranium above. The neck is a crucial area as a route of passage for major nerves, blood vessels, and organs passing between the head above and the trunk below. In addition to this, there are a few essential organs here too like the thyroid and parathyroid glands.

The neck is thinner than the head above, and the chest below. This is ideally suited to allow for the flexibility here and also to aid in the position of the head for maximizing the position of our sensory organs like our nose, eyes, ears, and mouth.

The neck is also an area crammed with many anatomical structures, and can at first appear daunting to understand its anatomy. Many organs, blood vessels, muscles, bones, and nerves are packed tightly into this space, and as such it makes the neck an area of high vulnerability. Indeed, organs like the trachea, esophagus, and the thyroid gland lack the bony protection that other sites in the body have the benefit of.

The neck, when viewed from the side, presents roughly a quadrilateral outline which has the following boundaries as highlighted in the following table (Table 3.1).

This area is divided by the sternocleidomastoid muscle into two triangles: an anterior in front of the muscle, and a posterior triangle behind this muscle.

3.1.1 Sternocleidomastoid

The sternocleidomastoid (where "cleido" means clavicle), or more simply put, sternomastoid, extends obliquely up the neck from the sternoclavicular joint to the mastoid process. It has two heads of origin. The

Essential Clinically Applied Anatomy of the Peripheral Nervous System in the Head and Neck.
DOI: http://dx.doi.org/10.1016/B978-0-12-803633-4.00003-X

Table 3.1 This Summarizes the Boundaries Which Limit the Triangles of the Neck

Borders of the triangles of the neck	
Superior	Inferior border of the mandible and a line drawn from the angle of the mandible to the mastoid process
Inferior	Superior aspect of the clavicle
Anterior	Anterior median line of the neck
Posterior	Anterior border of the trapezius

rounded, tendinous sternal head comes from the anterior aspect of the manubrium. The flatter clavicular head is attached to the superior aspect of the medial third of the clavicle. The clavicular head varies greatly in width, and a variable interval is found between the two heads. The muscle is inserted on the outer surface of the mastoid process and into the outer half of the superior nuchal line of the occipital bone.

The sternocleidomastoid is crossed by the platysma, external jugular vein, transverse cervical, and great auricular nerves. The muscle covers the great vessels of the neck, parts of other muscles (splenius, levator scapulae, digastric, sternothyroid, sternohyoid, and omohyoid), the cervical plexus, and the pleura. The triangles of the neck are therefore formed by this key muscle—the sternocleidomastoid as it is the common division of the triangles.

The sternocleidomastoid muscles acting together, that is both left and right sides, bend the head forward against resistance. Although the sternocleidomastoid muscles pull the cervical part of the vertebral column forward into flexion, their posterior fibers probably extend at the atlanto-occipital joints. The sternocleidomastoid muscles are active during extension at those joints. This muscle is also of importance in respiration only when the rate of ventilation is elevated and the ordinary muscles of inspiration are operating at a disadvantage.

When one of the sternocleidomastoid muscles contracts, the head is inclined laterally toward that side and the face rotated to the opposite side. In rotation without resistance, the sternocleidomastoids are usually active only toward the end of the movement.

3.1.2 Trapezius
The trapezius originates from the medial third of the superior nuchal line, external protuberance, ligamentum nuchae, and the spines of the

last cervical vertebrae, as well as the supraspinous ligament. Those fibers from the occipital bone and the ligamentum nuchae are inserted into the posterior border and upper surface of the lateral one-third of the clavicle. The other fibers are inserted onto the acromion and the spine of the scapula.

Both the trapezius and the sternocleidomastoid muscles are innervated by the accessory nerve, which is the 11th cranial nerve. There also comes an innervation from the ventral rami of the cervical nerves, which may give proprioceptive as well as motor innervation.

The upper fibers of the trapezius, alongside the levator scapulae, raise the shoulders and, acting with those of the opposite side, keep the shoulders braced by pulling the scapulae backwards. Their weakness would result in drooping of the shoulder. The middle and lower parts of the trapezius act alongside the rhomboids in retracting and steadying the scapula (squaring of the shoulders). The trapezius also has a role in rotating the scapula during abduction and elevation of the arm.

3.2 ANTERIOR TRIANGLE

The boundaries of the anterior triangle are as follows:

1. Superior—inferior border of the mandible and a line drawn from the angle of the mandible to the mastoid process
2. Anterior—Anterior median line of the neck
3. Posterior—anterior border of the sternocleidomastoid muscle

Within the anterior triangle of the neck, there are many anatomical structures crammed into a small space. These will be dealt with in this chapter only from the neurological perspective.

3.2.1 Cutaneous Nerves

Cutaneous innervation of the face overlaps that of the neck. Indeed, sensory innervation of the face arises from the terminal branches of the three main divisions of the trigeminal nerve that is the ophthalmic, maxillary, and mandibular divisions of this nerve. This has already been covered in the previous chapter, to which you should refer for further details about the trigeminal nerve, and its branches.

3.2.1.1 Ansa Cervicalis

The ansa cervicalis is a loop of nerves which are formed for the first three cervical nerves (C1–C3) and innervate the infrahyoid ("strap") muscles. The ansa cervicalis has a superior and inferior root.

The superior root of the ansa cervicalis is created by the first cervical nerve (C1). This root passes in the fibers of the hypoglossal nerve before coming away from it within the carotid triangle and thus forms its superior root. It passes round the occipital artery and passes down to the carotid sheath. It innervates the omohyoid muscle's superior belly as well as the superior portion of the sternothyroid and the sternohyoid muscles and unites with the inferior root of the ansa cervicalis.

The inferior root of the ansa cervicalis is formed by the unification of the second and third cervical nerves. It courses inferiorly on the outer aspect of the internal jugular vein, then crosses roughly midway down this vessel to pass anterior to this vein. It then passes in an anterior direction to unite with the superior root of the ansa cervicalis anterior to the common carotid artery. The inferior root of the ansa cervicalis may pass between the common carotid artery and the internal jugular vein. All the infrahyoid ("strap") muscles are innervated by the ansa cervicalis, except the thyrohyoid which is innervated by the first cervical nerve which accompanies the hypoglossal nerve for a short distance.

3.2.2 Cranial Nerves

There are three major cranial nerves which are present in the anterior triangles of the neck on the left and right sides—the glossopharyngeal, vagus, and hypoglossal nerves.

3.2.2.1 Glossopharyngeal Nerve

The ninth cranial nerve is the glossopharyngeal nerve. On emerging from the brainstem it passes through the cranial cavity to exit at the jugular foramen. It is primarily a sensory nerve but also contains motor and parasympathetic fibers. It supplies the posterior one-third of the tongue for taste and sensation, stylopharyngeus for its motor innervation, parasympathetic supply for the parotid gland, and receives sensory information from several neck structures as well as the ear. Clinically testing for the integrity of the glossopharyngeal nerve is simple. By asking the patient to say "ahhh" and testing the gag reflex should suffice. Isolated ninth nerve palsies are extremely rare, and they tend to occur when a pathology affects several cranial nerves found in close proximity like the facial, vagus, spinal accessory, and perhaps also the hypoglossal nerve. Further details of the glossopharyngeal nerve can be found in the previous chapter.

3.2.2.2 Vagus Nerve

The vagus nerve is the 10th cranial nerve. Like the glossopharyngeal nerve, again, it is a rather complex nerve having four nuclei and five different types of fibers in it. These convey information related to sensory, muscular activity, and autonomic functions. Its name comes from the Latin word vagary, meaning wandering. It has the longest course of the cranial nerves and is extensively distributed, especially below the level of the head. It contains the following types of fibers:

a. *Branchial motor*
   Supplying muscles of the pharynx and larynx
b. *Visceral sensory*
   This component of the vagus nerve is responsible for transmitting information from a wide variety of anatomical sites including the heart and lungs, pharynx and larynx, and the upper part of the gastrointestinal tract.
c. *Visceral motor*
   The visceral motor component carries parasympathetic fibers from the smooth muscle of the upper respiratory tract, heart, and gastrointestinal tract.
d. *Special sensory*
   The special sensation conveyed by the vagus nerve is for taste from the palate and epiglottis.
e. *General sensory*
   The general sensory component of the vagus nerve is concerned with information from parts of the ear and the dura within the posterior cranial fossa.

Further details of the vagus nerve can be found in the previous chapter.

3.2.2.3 Hypoglossal Nerve

The spinal accessory nerve is the 11th cranial nerve. It is a motor nerve (somatic motor) innervating two muscles—the sternocleidomastoid and trapezius. It has two components—a spinal part and a cranial part. The cranial part of the accessory nerve is from the vagus nerve. However, more recently, it has been shown that not all individuals may have a cranial root (Tubbs et al., 2014). However, when present (in the majority of cases), it joins with the spinal part of the accessory nerve for a short distance. The spinal part of the accessory nerve arises from the first five or six cervical spinal nerves. It has been shown that there is an elongated nucleus which extends from the first seven cervical vertebral levels, which provides the spinal portion of the accessory nerve (Pearson, 1937; Pearson et al., 1964). These branches arise from the lateral side of the spinal cord then form a nerve trunk. This spinal portion then ascends through the foramen magnum passing laterally to join with the cranial root.

As the two nerves join, they then pass through the jugular foramen briefly, along with the glossopharyngeal and vagus nerves. The cranial part then passes to the superior ganglion of the vagus. It then is distributed primarily in the branches of the vagus nerve, specifically the pharyngeal and recurrent laryngeal nerves.

The spinal portion then goes on to supply the sternocleidomastoid and trapezius in the neck. Further details of the hypoglossal nerve can be found in the previous chapter.

3.3 POSTERIOR TRIANGLE

The boundaries of the posterior triangle are as follows:

1. *Anterior—the posterior boundary of the sternocleidomastoid muscle*
2. *Posterior—the anterior border of the trapezius*
3. *Base—the middle one-third of the clavicle*

Within the posterior triangle of the neck, there are many anatomical structures crammed into a small space. These can be simply classified into the following categories of structures:

a. *Nerves*
   — Accessory nerve
   — Brachial plexus
   — Cervical plexus
   — Phrenic nerve

b. *Muscles*
   - Lower part of omohyoid
   - Scalene muscles (anterior, middle, and posterior)
   - Levator scapulae
   - Splenius
c. *Blood vessels*
   - Subclavian artery
   - Suprascapular artery
   - Transverse cervical artery
d. *Lymph nodes*

3.3.1 Cutaneous Nerves

In relation to the cervical plexus, this is a group of nerves that takes its origin from the ventral rami of the first four cervical nerves (C1–C4). This group of nerves is found deep to the sternocleidomastoid muscle and comprises of both sensory and motor nerves. The branches of the cervical plexus emerge from the posterior triangle of the neck at Erb's point. This point is found midway along the posterior border of the sternocleidomastoid muscle and approximately 2–3 cm superior to the clavicle.

As mentioned, the cervical plexus has cutaneous and muscular branches. The cutaneous branches are the lesser occipital nerve, transverse cutaneous (cervical) nerve of neck, great auricular nerve, and the supraclavicular nerves.

The muscular branches of the cervical plexus are as follows:

1. *Ansa cervicalis*
   The ansa cervicalis is a loop of nerves which are formed for the first three cervical nerves (C1–C3) and innervate the infrahyoid ("strap") muscles. The ansa cervicalis has a superior and inferior root.
   The superior root of the ansa cervicalis is created by the first cervical nerve (C1). This root passes in the fibers of the hypoglossal nerve before coming away from it within the carotid triangle and thus forms its superior root. It passes round the occipital artery and passes down to the carotid sheath. It innervates the omohyoid muscle's superior belly as well as the superior portion of the sternothyroid and the sternohyoid muscles and unites with the inferior root of the ansa cervicalis.

The inferior root of the ansa cervicalis is formed by the unification of the second and third cervical nerves. It courses inferiorly on the outer aspect of the internal jugular vein, then crosses roughly midway down this vessel to pass anterior to this vein. It then passes in an anterior direction to unite with the superior root of the ansa cervicalis anterior to the common carotid artery. The inferior root of the ansa cervicalis may pass between the common carotid artery and the internal jugular vein. All the infrahyoid ("strap") muscles are innervated by the ansa cervicalis, except the thyrohyoid which is innervated by the first cervical nerve which accompanies the hypoglossal nerve for a short distance.

2. *Phrenic nerve*

The phrenic nerve originates from cervical nerves three to five (ie, C3−5). It primarily has input from the fourth cervical nerve, but also can arise from the third and the fifth cervical nerves. The phrenic nerve is the sole motor innervation to the diaphragm. Arising at the upper region of the outside border of the anterior scalene muscle, the phrenic nerve courses downwards almost vertically across this muscle, posterior to the prevertebral fascia covering the front portion of this muscle.

The phrenic nerve descends through the neck deep to the sternocleidomastoid muscle, omohyoid's inferior belly, internal jugular vein, suprascapular and transverse cervical arteries, and on the left hand side, the thoracic duct. The nerve then runs anterior to the subclavian artery and posterior to the subclavian vein. It then enters the thoracic cavity by passing medially anterior to the internal thoracic artery. It passes anterior to the root of the lung, sandwiched between the mediastinal pleura and the fibrous pericardium. Accompanying the phrenic nerve at this point are the pericardiophrenic vessels.

On the right side, the phrenic nerve is more vertical and a little shorter than the left. It is separated in the root of the neck from the subclavian artery by the anterior scalene muscle. The phrenic nerve on the right side then is found lateral to the brachiocephalic vein, superior vena cava, and the fibrous pericardium.

On the left side, the phrenic nerve approximates the route taken by the right phrenic nerve in the root of the neck. On its descent, the phrenic nerve then crosses over the internal thoracic arytery closely related to the medial apex of the lung, close to the subclavian and common carotid arteries. It then passes medially and anteriorly superficial to the vagus nerve superior to the aortic arch and posterior to the brachiocephalic vein.

3. Branches innervating the anterior and middle scalene muscles.

3.3.1.1 Lesser Occipital Nerve

The lesser occipital nerve originates from the second and third cervical spinal nerves, from the cervical plexus. It runs along the posterior border of the sternocleidomastoid muscle, and then passes to the rear of the auricle. As such, it innervates the scalp behind the auricle.

3.3.1.2 Transverse Cutaneous (Cervical) Nerve of Neck

The transverse cutaneous, or transverse cervical nerve of the neck arises from the second and third cervical vertebral nerves. It turns around the middle of the posterior border of the sternocleidomastoid muscle, running obliquely anteriorly deep to the external jugular vein, and crosses that muscle deep to the platysma. Under the platysma it then divides into branches which innervate the skin on the side and the anterior aspect of the neck.

3.3.1.3 Great Auricular Nerve

The cervical plexus is located deeply in the upper part of the neck, under cover of the internal jugular vein and the sternocleidomastoid muscle. It is formed by the ventral rami of the first four cervical nerves. The superficial cutaneous branches of the plexus are the lesser occipital, great auricular, transverse cervical nerves, and the supraclavicular nerves. All of these branches emerge near the middle of the posterior border of the sternocleidomastoid muscle.

The great auricular nerve is from the cervical plexus as a superficial branch, originating from the second and third cervical vertebral levels (C2−C3). The great auricular nerve exits the cervical plexus at the posterior aspect of the sternocleidomastoid muscle at a point called Erb's point. Erb's point is where some of the cervical nerves converge and can also be where the great auricular and accessory nerves are encountered. It is defined as "a circumscribed point, about 2−3 cm above the clavicle, somewhat outside of the posterior border of the sternomastoid and immediately in front of the sixth cervical vertebra" (Landers and Maino, 2012). It can also be thought of as being found along the posterior border of the sternocleidomastoid at about 6.5 cm inferior to the external auditory canal as a 3 cm circle there (Gloster, 2008).

The great auricular nerve then ascends obliquely across the sterno-cleidomastoid muscle to supply the skin over the parotid gland and over the mastoid process, together with both surfaces of the auricle.

3.3.1.4 Supraclavicular Nerves

The supraclavicular nerves originate from the third and fourth cervical vertebral levels and emerge as a common trunk under the cover of the sternocleidomastoid muscle. The trunk divides into anterior, middle, and posterior supraclavicular nerves, which descend under the cover of the platysma in the posterior triangle, cross the clavicle superficially, and suppy skin over the shoulder as far forward as the median plane.

Nathe et al. (2011) had undertaken a cadaveric study where they examined the distribution of the supraclavicular nerves to inform surgeons who may want to treat clavicular shaft fractures by surgery, to aid function and normal anatomical shape. They had shown that there are two "safe zones" for surgical access to the clavicle for nonunion or malunion surgery. They had shown that there is no medial nerve branching within 2.7 cm of the sternoclavicular joint, and no lateral nerve was present within 1.9 cm of the acromioclavicular joint.

3.4 CLINICAL APPLICATIONS

3.4.1 Clinical Testing of the Supraclavicular Nerves

The testing of the supraclavicular nerves can be undertaken during a sensory examination, and as such can be tested as a dermatome. Each spinal nerve has a distribution called a *dermatome*. This is the area that is supplied by a single spinal nerve by its sensory fibers running in its dorsal root. However, sectioning of a single spinal nerve rarely will result in complete loss of sensation, or anesthesia. This is because other adjacent spinal nerves will also be carrying fibers from that site. Therefore the more likely scenario is a reduced level of sensation, or hypoesthesia.

There are many versions of *dermatome maps* which can be used clinically to demonstrate the site(s) of pathology of a patient with a suspected spinal nerve lesion. The most common one to be used clinically is the American Spinal Injury Association's (ASIA) worksheet produced as the International Standards for Neurological Classification of Spinal Cord Injury (ISNCSCI).

3.4.2 Fine Touch

Tip!

When examining a patient's sensory system, do not provide suggestions to them as it may influence their interpretation of the examination. DO NOT say to the patient if they notice any changes during the examination, as this suggests that they should expect to notice a change.

3.4.3 Light Touch Examination

The purpose of this is to serve as an introductory examination to identify any areas or regions where there may be pathologies present. Simple testing should first be done and as the extremities are easy to access without causing too much discomfort to the patient, these should be examined first.

Tip!

Testing of *light touch sensation* to identify any areas of abnormality MUST be done when the patient's eyes are closed. This ensures that they do not anticipate where the stimulus is applied. Equally, you MUST tell the patient what you are going to do before doing it. This helps build trust with the patient but also allows them to be fully informed about what you are doing, and that they consent to it.

a. Introduce yourself to the patient stating who you are, and in what capacity and grade you are functioning for example student, physician, surgeon, therapist etc.
b. Advise the patient that you want to test for sensation in their extremities initially when their eyes are closed.
c. Tell them that you will touch them with either cotton wool, or using a specialist Von Frey filament on various regions of the arms, hands, legs, and feet on both the left and right hand sides. You do not need to use the term Von Frey but inform them that it is material made of nylon that allows you to test sensation.
d. Ask the patient if they have understood what you have said and answer any questions they may have.
e. Now, ask the patient to close their eyes and to say "yes" every time they feel you touch, and also to report any feelings of discomfort, pain, or abnormal sensations.

f. Working systematically with the cotton wool or Von Frey fibers, touch all dermatome regions of the upper and lower limbs on both the right and left sides of the limbs.

g. Try to compare both the left and right hand sides of each dermatome region.

Tip!

A dermatome is a region of skin supplied by one single spinal nerve. It allows you to determine what spinal level pathology may be at. The distribution of our dermatomes can be slightly variable but can give an indication at what level approximately the pathology may exist for example upper cervical, lower thoracic, and so on.

If you identify a level, or levels, where there appears to be an alteration in sensation, identify exactly at what point this occurs. When recording the findings from the clinical examination, it can be used on the American Spinal Injury Association's (ASIA) worksheet produced as the International Standards for Neurological Classification of Spinal Cord Injury (ISNCSCI).

3.4.4 International Standards for Neurological Classification of Spinal Cord Injury (ISNCSCI)

This sheet allows for assessment and classification of motor and sensory function and is classified as follows:

3.4.4.1 Sensory Assessment

Light touch and pin-prick are assessed separately and a score out of 2 is given for each dermatome on the right and left side of the body.

The dermatomes that are assessed are cervical (C2, C3, C4, C5, C6, C7, C8), thoracic (T1, T2, T3, T4, T5, T6, T7, T8, T9, T10, T11, T12), lumbar (L1, L2, L3, L4, L5), and sacral (S1, S2, S3, S4–S5).

For sensation the following scoring system is used:

Sensory Score	Classification
0	Absent
1	Altered
2	Normal
NT	Not testable

A score is given for light touch right (*LTR* totaling 56 ie 2 for each of the vertebral levels stated above) and light touch left (*LTL* totaling 56 ie 2 for each of the vertebral levels stated above). This is recorded as follows:

$$LTR + LTL = \text{out of } 112$$

A score is also given for pin-prick right (*PPR* totaling 56 ie 2 for each of the vertebral levels stated above) and pin-prick left (*PPL* totaling 56 ie 2 for each of the vertebral levels stated above). This is recorded as follows:

$$PPR + PPL = \text{out of } 112$$

There is a corresponding system for scoring the motor system, and the power of muscles and presence, or otherwise, of reflexes.

3.5 CLINICAL APPLICATIONS

3.5.1 Injury to the Supraclavicular Nerves

Due to the superficial location of the supraclavicular nerves, it makes them vulnerable to injury, especially during a fracture of the clavicle, including displaced and comminuted fractures from high impact trauma (Jeray, 2007). Indeed, the clinician should keep in mind that sensation may be altered in a slightly wider area than initially expected due to the variability in distribution of dermatomes. Symptoms of nerve damage for example pain, hypoesthesia, or parasthesia may occur not only in the expected distribution of the supraclavicular nerves but also along the proximal deltoid and the posterolateral region of the scapular belt (Gelberman et al., 1975).

If surgical access to the clavicle is required for example to realign the clavicle and perhaps to insert surgical plates, variability of parasthesia is rather large in these patients requiring plate insertion. The Canadian Orthopedic Trauma Society (2007) and Shen et al. (1999) stated that incidence of parasthesia on patient's requiring plate insertion and who experience parasthesia as a result of the procedure varied from 12–29%.

3.5.2 Supraclavicular Nerve Block

Blockade of the brachial plexus can be achieved through anaesthetizing the supraclavicular nerves which is where the trunks of C5–T1 (vertebral level of origin of the brachial plexus) are most compact. Any surgery of the upper limb, really from mid-humerus level down can be achieved by blockade either of the supraclavicular, interscalene, transcalene, axillary, and infraclavicular approaches (Nguyen et al., 2007). However, the lateral approach to the supraclavicular nerve, especially under ultrasound guidance has been the most favored approach for anesthetizing the nerves of the upper limb (Hempel et al., 1981; Kothari, 2003). However, complications can arise like vessel puncture 9 with the nearby subclavian artery and also pneumothorax.

Indeed, blockade of the supraclavicular nerves can also be undertaken for more chronic pain, for example like injury to the rotator cuff muscles, tendinitis, capsulitis, or rheumatoid arthritis at or close to the site of these nerves (Allen et al., 2010; Fernandes and Fernandes, 2010).

3.5.3 Cranial Nerves
The cranial nerve which is present in the posterior triangle of the neck is that of the accessory nerve.

3.5.3.1 Accessory Nerve
The spinal accessory nerve is the eleventh cranial nerve. It is a motor nerve (somatic motor) innervating two muscles—the sternocleidomastoid and trapezius. It has two components—a spinal part and a cranial part. The cranial part of the accessory nerve is from the vagus nerve. However, more recently, it has been shown that not all individuals may have a cranial root (Tubbs et al., 2014). However, when present (in the majority of cases), it joins with the spinal part of the accessory nerve for a short distance. The spinal part of the accessory nerve arises from the first five or six cervical spinal nerves. It has been shown that there is an elongated nucleus which extends from the first seven cervical vertebral levels, which provides the spinal portion of the accessory nerve (Pearson, 1937; Pearson et al., 1964). These branches arise from the lateral side of the spinal cord then form a nerve trunk. This spinal portion then ascends through the foramen magnum passing laterally to join with the cranial root.

As the two nerves join, they then pass through the jugular foramen briefly, along with the glossopharyngeal and vagus nerves. The cranial part then passes to the superior ganglion of the vagus. It then is distributed primarily in the branches of the vagus nerve, specifically the pharyngeal and recurrent laryngeal nerves.

The spinal portion then goes on to supply the sternocleidomastoid and trapezius in the neck.

The accessory nerve has two roots—a cranial and spinal division. The cranial root arises from the inferior end of the nucleus ambiguus and perhaps also from the dorsal nucleus of the vagus nucleus. The fibers of the nucleus ambiguus are connected bilaterally with the corticobulbar tract (motor neurons of the cranial nerves connecting the cerebral cortex with the brainstem). The cranial part leaves the medulla oblongata as four or five rootlets uniting together, and then to join with the spinal part of the accessory nerve just as it enters the jugular foramen. At that point, it can send occasional fibers to the spinal part. It is only united with the spinal part of the accessory nerve for a brief time before uniting with the inferior ganglion of the vagus nerve. These cranial fibers will then pass to the recurrent laryngeal and pharyngeal branches of the vagus nerve, ultimately destined for the muscles of the soft palate (not tensor veli palatini (supplied by the medial pterygoid nerve of the mandibular nerve)).

The spinal root arises from the spinal nucleus found in the ventral gray column extending down to the fifth cervical vertebral level. These fibers then emerge from the spinal cord arising from between the ventral and dorsal roots. It then ascends between the dorsal roots of the spinal nerves entering the cranial cavity through the foramen magnum posterior to the vertebral arteries. It then passes to the jugular foramen, where it may receive some fibers from the cranial root. As it then exits the jugular foramen, it is closely related to the internal jugular vein. It then courses inferiorly passing medial to the styloid process and attached stylohyoid. It also is found medial to the posterior belly of digastric. The spinal root then supplies the sternocleidomastoid muscle on its medial aspect.

The cranial root then enters the posterior triangle on the neck lying on the surface of the levator scapulae at approximately midway down the sternocleidomastoid. As it passes inferiorly through the posterior triangle of the neck, and just above the clavicle, it then enters the trapezius muscle on its deep surface at its anterior border. The third and fourth cervical vertebral spinal nerves also supply the trapezius forming a plexus of nerves on its deeper surface.

3.6 CLINICAL APPLICATIONS

3.6.1 Clinical Testing of the Accessory Nerve
3.6.1.1 Testing at the Bedside

From the clinical perspective, the accessory nerve supplies the *sterno-cleidomastoid* and *trapezius* muscles, and as such, it is those that are tested when assessing the integrity of the nerve.

The sternocleidomastoid muscle has two functions dependent on whether it is acting on its own, or with the opposite side. If the sterno-cleidomastoid is acting on its own, it tilts the head to that side it contracts and, due to its attachments and orientation, rotates the head so that the face moves in the direction of the opposite side. Therefore, if the left sternocleidomastoid muscle contracts, the face turns to the right hand side, and vice versa.

If, however, both sternocleidomastoid muscles contract, the neck flexes and the sternum is raised, as in forced inspiration.

The trapezius is an extremely large superficial muscle of the back. It is comprised of three united parts—superior, middle, and inferior. It is involved in two main functions dependent on if the scapula or the spine is stable. If the spinal part is stable, it helps to move the scapula, and if the scapula is stable, it helps to move the spine. Trapezius is involved in a variety of movements. The upper fibers raise the scapula, the middle fibers pull the scapula medially, and the lower fibers move the medial side of the scapula down. Therefore, the trapezius is involved in both elevation and depression of the scapula, dependent on which part is contracting. As well as this, the trapezius also rotates and retracts the scapula.

Testing of the accessory nerve is done as follows:

1. ALWAYS inform the patient of what you will be doing, after introducing yourself and taking a detailed clinical history.
2. When examining a patient, ensure that you just observe the patient and try to identify if there is any obvious deformity, or asymmetry of the shoulder and neck region. It may be that you will see an obvious weakness or asymmetrical position of the patient's neck and/or upper limbs.
3. First, you can assess the sternocleidomastoid.

4. You can ask the patient to rotate their head to look to the left and right hand sides to identify any obvious abnormality.
5. Then, ask the patient to look to one side and test the muscle against resistance.
6. For example, if the patient looks to the right side, place the ball of your hand on their left mandible.
7. Ask the patient to press into your hand.
8. Repeat this on the opposite side.
9. Then, you need to assess the trapezius.
10. First you can ask the patient to raise their shoulders, as in shrugging.
11. Observe any gross abnormality.
12. Then while the patient is raising their shoulders, gently press down on them as they lift their shoulders.
13. Assess any weakness which may be present, noting which side is affected.

Tip!

When assessing the function of the sternocleidomastoid and trapezius, it may help in examining the unaffected side first, especially if the patient complains of pain or discomfort on one side. This helps build up trust with the patient, but also minimizes causing them any pain or discomfort.

Tip!

Weakness in rotating the head to the left hand side, when examining sternocleidomastoid, suggests a pathology with the right accessory nerve, and vice versa.

3.6.1.2 Advanced Testing

The bedside testing should be sufficient in examining the accessory nerve, but the results of this may help direct toward any specialist investigations that may be relevant for the patient. Electromyography (EMG) studies may be deemed relevant, as well as further cranial and neck investigation by means of CT and /or MRI scanning, dependent on the clinical history and physical examination findings.

3.7 CERVICAL NERVES AND BRACHIAL PLEXUS

The cervical plexus is located deeply in the upper part of the neck, under cover of the internal jugular vein and the sternocleidomastoid muscle. It is formed by the ventral rami of the first four cervical nerves. The superficial cutaneous branches of the plexus are the lesser occipital, great auricular, transverse cervical nerves, and the supraclavicular nerves. All of these branches emerge near the middle of the posterior border of the sternocleidomastoid muscle.

The superficial branches of the cervical plexus are as follows:

a. *Lesser occipital nerve*
   The lesser occipital nerve hooks around the accessory nerve, ascends along the posterior border of the sternocleidomastoid muscle, runs posterior to the auricle, and innervates the skin on the side of the head and on the cranial surface of the auricle.
b. *Great auricular nerve*
   The great auricular nerve is from the cervical plexus as a superficial branch, originating from the second and third cervical vertebral levels (C2–C3). The great auricular nerve exits the cervical plexus at the posterior aspect of the sternocleidomastoid muscle at a point called Erb's point.
c. *Transverse cervical nerve*
   The transverse cervical nerve turns around the middle of the posterior border of the sternocleidomastoid muscle and crosses that muscle deep to the platysma. It divides into branches which innervate the skin on the side and anterior aspect of the neck.
d. *Supraclavicular nerves*
   The supraclavicular nerves originate from the third and fourth cervical vertebral levels and emerge as a common trunk under the cover of the sternocleidomastoid muscle. The trunk divides into anterior, middle, and posterior supraclavicular nerves, which descend under the cover of the platysma in the posterior triangle, cross the clavicle superficially, and supply skin over the shoulder as far forward as the median plane.

3.7.1 Brachial Plexus

The nerves to the upper limb arise from the brachial plexus, which is a large and very important structure located partly in the neck and partly in the axilla. The brachial plexus is formed by the union of the ventral rami of the lower cervical nerves (ie, C5, C6, C7, C8) and the greater part of the ventral ramus of the first thoracic nerve (ie, T1). Therefore, it can be written that the origin of the brachial plexus arises from C5–T1. However, the brachial plexus frequently can receive contribution from the fourth cervical nerve (C4) above, or the second thoracic nerve (T2) below.

When the fourth cervical nerve is large and the first thoracic nerve is small, the plexus is described as being *prefixed*. If there is more contribution from the first and second thoracic nerves, the brachial plexus is described as being postfixed. If the first rib is rudimentary, the second thoracic nerve can provide the brachial plexus with a larger innervation.

Following on from its origin, the brachial plexus then descends into the lower part of the neck called the posterior triangle. The posterior triangle of the neck has the following boundaries:

1. *Anterior*—the posterior border of the sternocleidomastoid muscle
2. *Posterior*—the anterior border of the trapezius muscle
3. *Base*—the middle third of the clavicle
4. *Apex*—the sternocleidomastoid and the trapezius, at the superior nuchal line of the occipital bone
5. *Roof*—the superficial layer from the deep cervical fascia

The brachial plexus is found above the clavicle, and posterior and lateral to the sternocleidomastoid muscle. It is posterosuperior to the third part of the subclavian artery, and is crossed by the lower belly of the omohyoid muscle. In terms of surface anatomy, the brachial plexus can be felt in the living both superior and inferior to the omohyoid, located in the angle between the clavicle and the sternocleidomastoid. From the surface anatomy perspective, the brachial plexus can be found in the neck below a line drawn from the posterior margin of the sternocleidomastoid at the level of the cricoid cartilage to the midpoint of the clavicle.

Superficial to the brachial plexus within the neck are several structures namely the supraclavicular nerves, platysma, external jugular vein, inferior belly of omohyoid, and the descending scapular and transverse cervical arteries.

The brachial plexus then passes posterior to the medial two-thirds of the clavicle, running with the axillary artery as it passes deep to the pectoralis major. The cords of the brachial plexus are arranged around the axillary artery posterior to the pectoralis minor. The brachial plexus cords are held together with the axillary artery with the axillary sheath. The terminal branches are then given off at the lower outer border of the pectoralis minor.

The ventral rami of the brachial plexus from the fifth and sixth cervical nerve (C5−6) form the *upper trunk*. The *middle trunk* originates from the seventh cervical nerve (C7). The *lower trunk* of the brachial plexus is formed from the eight cervical and the first thoracic spinal nerves (C8−T1). Each of these trunks splits into *anterior* and *posterior divisions*. The division that happens at this point gives an idea of what nerves will innervate the front, and which will innervate the back.

The *lateral cord* is formed from the *anterior divisions* of the *upper* and *middle trunks*. The *medial cord* is from the *anterior division* of the *lower trunk*. The *posterior cord* arises from the *three posterior divisions*. The lateral, medial, and posterior cords tend to lie posterior to the axillary artery, and then wind round the artery to lie either lateral, medial, or posterior relative to this vessel. At the lower outer border of the pectoralis minor, the lateral, medial, and posterior cords split into their terminal branches. Due to the complex formation of the brachial plexus, the terminal branches tend to have several spinal nerves contributing to each of the final branches. Table 3.2 provides a broad outline of where each region is found anatomically. Table 3.3 provides an easy summary to identify what contributes to each peripheral (terminal) nerve from the roots, trunks, divisions, and cords.

Table 3.2 This Table Summarizes the Anatomical Location of Each Part of the Brachial Plexus (ie, Roots (Ventral Rami), Trunks, Divisions, Cords, and Terminal Branches)

Part of brachial plexus	Location
Roots and trunks	Neck, close to the subclavian artery
Divisions	Posterior to the clavicle
Cords and terminal branches	Axilla, close to the axillary artery

Table 3.3 This Table Summarizes What Forms the Terminal Branches (ie, Peripheral Nerves). It Also Highlights What Forms the Cords, Divisions, and Trunks, from the Ventral Rami of C5–T1

Roots	Trunks	Divisions	Cords	Terminal branches
C5 C6	Upper	Anterior Posterior	Lateral	Musculocutaneous
C7	Middle	Anterior Posterior	Medial	Median Ulnar
C8 T1	Lower	Anterior Posterior	Posterior	Radial Axillary

Note: The branches of the ventral rami of C5–T1 forming the brachial plexus include the dorsal scapular nerve, long thoracic nerve, and also small branches to the scalene and longus colli muscles.

3.7.2 Clinical Testing of the Brachial Plexus

For the purposes of this, the terminal branches as the final points of termination of the brachial plexus will be discussed here. In other words the following nerves will be dealt with in clinical examination of the brachial plexus:

1. Musculocutaneous nerve
2. Median nerve
3. Ulnar nerve
4. Radial nerve
5. Axillary nerve

For further detail of the other branches of the brachial plexus, and clinical examination and related pathologies, please refer to the companion text to this book entitled *Essential Clinically Applied Anatomy of the Peripheral Nervous System in the Limbs* (Rea, 2015a,b).

3.7.3 Musculocutaneous Nerve

The musculocutaneous nerve originates from the fifth to seventh cervical nerves (C5–7), typically from the lateral cord and then pierces the coracobrachialis (Figure 3.1). It sometimes can also convey some or all of the lateral head of the median nerve and give these fibers to the medial head of the medial nerve with a communication in the arm. Therefore, the lateral cord can divide lower than usual. In some cases, the musculocutaneous nerve can travel with the lateral head of the median nerve and subsequently be given back as a

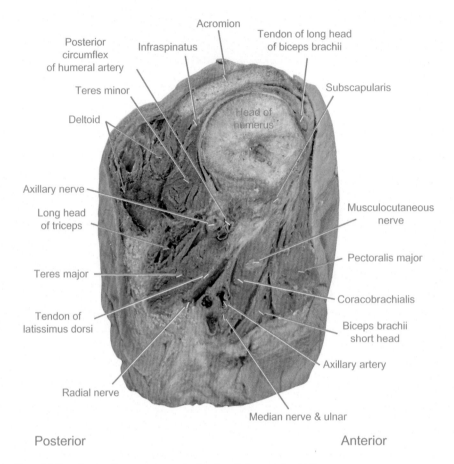

Figure 3.1 Note the musculocutaneous nerve passing through the coracobrachialis.

communication to the musculocutaneous nerve. Ultimately, the musculocutaneous nerve innervates the flexor muscles (described later) on
the anterior aspect of the arm, skin on the lateral side of the forearm,
and also to the elbow joint. The innervation of coracobrachialis can
come from the lateral cord separately, rather than from the musculocutaneous nerve itself.

If the musculocutaneous nerve arises from within the axilla, it usually pierces the coracobrachialis. It will then pass inferiorly between
the biceps superficially and the brachialis deeper, thus reaching the lateral aspect of the arm.

Table 3.4 This Summarizes the Origin, Insertion, and Functions of Each of the Muscles Supplied by the Musculocutaneous Nerve

Muscle	Origin	Insertion	Function
Coracobrachialis	Coracoid process	Middle third of medial side of humerus	Flexion of arm Adduction of humerus
Biceps brachii	Short head – coracoid process of scapula Long head – supraglenoid tubercle	Radial tuberosity Bicipital aponeurosis	Flexion of the elbow Flexion of the shoulder Abduction of the shoulder Supination at the radioulnar joint
Brachilais	Distal two-thirds of anteromedial and anterolateral surfaces of the humerus	Capsule of the elbow joint and anterior surface of the coronoid process and tuberosity of the ulna	Flexion of elbow

The musculocutaneous nerve innervates the coracobrachialis, biceps, and the brachialis as well as the elbow joint (see Table 3.4). It will then become the *lateral antebrachial cutaneous nerve*, or the *lateral cutaneous nerve of forearm*, which then goes through the fascia lateral to the tendon of the biceps tendon, just superior to the elbow. This will then divide into an *anterior (volar)* and *posterior (dorsal)* branch which may lie posterior to the cephalic vein. The anterior branch will innervate the skin on the anterior aspect of the radial side of the forearm as far as the thenar eminence. The posterior branch however innervates the skin on the lateral and posterior and lateral region of the forearm as far as the wrist. These branches will innervate a variable amount of skin on the dorsal aspect of the hand.

3.8 CLINICAL APPLICATIONS

When undertaking any clinical history taking or examination, you should always do the following, and follow a logical and systematic format:

a. Introduce yourself to the patient
b. Advise them of what position you hold for example student, specialty grade, consultant etc.
c. Your reason for consulting with them, or to find out why they have presented to you
d. Always take a thorough and detailed history, which will be guided by the presenting signs and symptoms

e. When examining the patient, always tell them what you will ask them to do, or what region of the body you will be examining, with specific instructions and ensure they give consent

A detailed examination and history taking should be completed as described in Chapter 1. Examination for the functioning of the biceps brachii, brachioradialis, and the coracobrachialis should be done by determining the power and functioning of the range of movements as follows:

3.8.1 Coracobrachialis

Assessment of the shoulder joints should be performed as detailed in the previous chapter. In addition to this, the elbow joint should be assessed as follows:

1. *Inspection*
   Inspection of the elbow joints should be done by standing behind the patient and observing the elbows in the fully extended position. Note any abnormality or asymmetry.
2. *Palpation*
   When feeling the elbow joint, be careful not to hurt the patient, especially if there is localized tenderness. Identify any nodules, points of tenderness, or any obvious signs of inflammation.
3. *Movement*
   a. The range of flexion-extension and pronation-supination should be assessed with active movements.
   b. With the elbows flexed at 90° and elbows by the patient's side, pronation and supination should be assessed. If this is not done, abduction and rotation of the shoulder can mislead the examiner for the appearance of pronation and supination.
   c. The examiner should gently palpate the elbow joint during flexion and extension, and also over the head of the radius during pronation and supination to identify if crepitus is present.

For completeness in the examination of the elbow joint, you must test for epicondylitis.

a. You must ask the patient to tightly grip their hand when the elbow is fully extended and also when partly flexed. If the patient has tennis elbow, the former maneuver will be painful, and the second not so much, or not at all. If the tennis elbow is severe, the patient may actually find it difficult for simply extending the elbow.

b. Then, place the patient's arm into the "waiter's tip" position that is full elbow extension, pronation of the forearm and wrist flexed. Them, ask the patient to attempt to extend their fingers against resistance. This will result in exacerbation of the pain at the insertion point of the extensors.

c. In assessing medial epicondylitis, ask the patient to put their arm in the extended position and supinate their forearm. Then, ask the patient to attempt to flex their wrist or fingers against resistance. This will result in pain at the flexor insertion, and can be useful in assessing medial epicondylitis.

3.8.2 Median Nerve

The median nerve originates from primarily the sixth cervical nerve to the first thoracic nerve that is C6–T1, though sometimes can have input also from the fifth cervical nerve. Therefore, the median nerve can originate from C5–T1.The median nerve is formed on the lateral aspect of the axillary artery by heads derived from the medial and lateral cords from the brachial plexus. It then continues on the lateral side of the brachial artery. At approximately the middle of the arm, the median nerve passes anterior to the brachial artery, but sometimes courses posterior to this vessel.

The median nerve then courses medial to the brachial artery (Figure 3.2). In the cubital fossa, the median nerve lies behind the median cubital vein and under the bicipital aponeurosis, providing a branch to the elbow joint. The nerve then leaves the cubital fossa typically by passing between the two heads of pronator teres. It is separated from the ulnar artery by the deep or ulnar head of that muscle.

The nerve then courses down the middle of the forearm to the midpoint between the styloid processes. Just prior to passing deep to the flexor retinaculum, it lies superficial between the tendons of the flexor carpi radialis and the palmaris longus. The median nerve then passes deep to the tendinous arch connecting the two heads of flexor digitorum superficialis. The nerve generally stays under the flexor digitorum superficialis and on the flexor digitorum profundus until it reaches the wrist joint.

At the wrist, the median nerve becomes superficial by passing between the flexor carpi radialis and the flexor digitorum superficialis. When present, the palmaris longus partly covers it. The median nerve

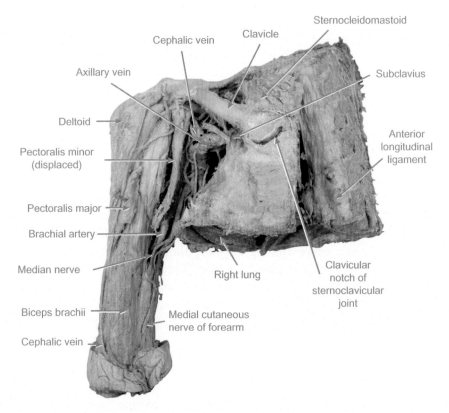

Figure 3.2 This shows the location of the median nerve within the arm.

then enters the hand by passing through the carpal tunnel behind the flexor retinaculum but in front of the flexor tendons. The median and ulnar nerves may communicate with each other in the forearm, and this typically occurs over the flexor digitorum profundus.

3.8.3 Branches of the Median Nerve
The median nerve gives off a variety of branches in its journey through the arm and forearm to the hand. These are summarized in the following table (Table 3.5).

3.9 CARPAL TUNNEL

The carpal tunnel is an osteofibrous canal situated at the wrist joint. The carpal tunnel is bounded by the carpal bones deeply, thus forming its floor, and the flexor retinaculum superficially, forming its roof. The

Table 3.5 This Table Summarizes the Innervation of Structures by the Median Nerve, Both Motor and Sensory

Median nerve branches		
FOREARM		
Muscular branches	Pronator teres Flexor carpi radialis Palmaris longus Flexor digitorum superficialis	
Anterior interosseous nerve	Flexor digitorum profundus (lateral part)	
	Flexor pollicis longus	
Palmar cutaneous branch	Lateral aspect of the palm (not digits)	
HAND		
Lateral division	Muscular branch	Abductor pollicis brevis
		Flexor pollicis brevis
		Opponens pollicis
	Palmar digital nerves	All of the thumb
		Lateral side of the index finger
Medial division	Medial side of index finger and middle finger, and lateral aspect of the ring finger	
	First and second lumbricals	

flexor retinaculum attaches onto the pisiform, tuberosity of the scaphoid and the trapezium and the hook of the hamate bone. It is divided into two layers at the radial side due to the tendon of the flexor carpi radialis. It therefore contains a deep and superficial layer. The carpal tunnel contains the median nerve as well as nine tendons. Therefore, the carpal tunnel contains the following:

1. Median nerve and tendons of:
2. Flexor digitorum profundus (x4)
3. Flexor digitorum superficialis (x4)
4. Flexor pollicis longus

In close contact with the carpal tunnel, though not passing through it are the tendons of the following muscles:

1. Flexor carpi ulnaris
2. Flexor carpi radialis
3. Palmaris longus

The following table (Table 3.6) summarizes the origin, insertion point, functions, and innervations of those muscles described.

Table 3.6 This Table Summarizes the Muscles Innervated by the Median Nerve, and Related Muscles and Nerves

Muscle	Origin	Insertion	Innervation	Function
Flexor digitorum profundus	Anterior surface of ulna (upper 2/3 to 3/4) Coronoid process Interosseous membrane	Bases of distal phalanges of the fingers	Medial: Ulnar nerve Lateral: Anterior interosseous nerve (from median nerve)	Flexes distal phalanges, typically with flexion of the middle phalanges by the flexor digitorum superficialis
Flexor digitorum superficialis	Common tendon from the medial epicondyle of humerus Anterior superior border of radius	Middle phalanges of the four fingers, anteriorly	Median nerve	Flexion of the fingers, mainly at the proximal interphalangeal joints
Flexor pollicis longus	Anterior surface of radius (upper 2/3 to 3/4) Interosseous membrane	Base of distal phalanx (palmar surface)	Anterior interosseous nerve (from median nerve)	Flexion of distal phalanx of the thumb
Flexor carpi ulnaris	Common tendon from the medial epicondyle of humerus Medial surface of the olecranon	Pisiform Hook of hamate and the base of the fifth metacarpal (via the pisohamate and the pisometacarpal ligaments)	Ulnar nerve	Flexion of the hand Adduction of the hand, with extensor carpi ulnaris Steadies pisiform during abduction of the little finger by abductor digiti minimi Synergistic action with flexor carpi radialis steadying the wrist joint during extension of the fingers Steadies hand during extension and abduction of the thumb, working alongside the extensor carpi ulnaris
Flexor carpi radialis	Common tendon from the medial epicondyle of humerus	Anterior aspect of the base of the second and third metacarpal bones	Median nerve	Flexion of the hand Along with the radial extensors, it aids abduction of the hand Synergistic muscle along with flexor carpi ulnaris, it steadies the wrist during extension of the fingers
Palmaris longus	Common tendon from the medial epicondyle of humerus	Flexor retinaculum Apex of palmar aponeurosis	Median nerve	Tenses the palmar aponeurosis on movement of the hand Weak flexor of the hand

3.10 CLINICAL APPLICATIONS

When undertaking any clinical history taking or examination, you should always do the following, and follow a logical and systematic format:

a. Introduce yourself to the patient
b. Advise them of what position you hold for example student, specialty grade, consultant etc.
c. Your reason for consulting with them, or to find out why they have presented to you
d. Always take a thorough and detailed history, which will be guided by the presenting signs and symptoms
e. When examining the patient, always tell them what you will ask them to do, or what region of the body you will be examining, with specific instructions and ensure they give consent

The following table (Table 3.7) summarizes the nerve roots, muscles supplied, and relevant clinical tests to ask the patient to perform in assessing the median nerve.

As well as testing the integrity of the muscles innervated by the median nerve, it is also essential to assess the sensory distribution of the median nerve that is all of the thumb, lateral side of the index

Table 3.7 This Summarizes the Ways to Clinically Examine the Media Nerve, and What to Ask the Patient to Do, and What Each Muscle Is Being Assessed For		
Nerve root	Muscle	Ask the patient to do...
C6-7	Pronator teres	Keep the arm pronated against resistance from the examiner
C6-8	Flexor carpi radialis	Flexion of the wrist to the radial side
C7-T1	Flexor digitorum superficialis	Resist extension to the proximal interphalangeal joints. The proximal phalanges should be fixed
C8-T1	Flexor digitorum profundus (1st and 2nd)	Resisting extension of the distal interphalangeal joints
C8-T1	Flexor pollicis longus	Resisting extension of the thumb at the interphalangeal joints. The proximal phalanges should be fixed
C8-T1	Abductor pollicis brevis	Thumb abduction
C8-T1	Opponens pollicis	Opposition of the thumb i.e. the patient has to bring the thumb to touch the 5th finger tip
C8-T1	1st and 2nd lumbricals	Extension of the proximal interphalangeal joint against resistance by the examiner with the metacarpophalangeal joint positioned hyper-extended

finger, medial side of the index finger and middle finger, and the lateral aspect of the ring finger. This should be documented according to the International Standards for Neurological Classification of Spinal Cord Injury (ISNCSCI).

3.11 CLINICAL APPLICATIONS

Perhaps the most common, and widely appreciated condition to affect the upper limb nerves is carpal tunnel syndrome. This is where the median nerve is compressed in the carpal tunnel, as previously described. It results in the patient having numbness and parasthesia in the fingers, typically affecting the thumb, index, and middle fingers, though some patients may feel as though their whole hand is affected. Early in the course of this pathology, the symptoms tend to be worse at night and tend to disappear by the morning. If there is progressive worsening of carpal tunnel syndrome, it can result in muscle wasting of the thenar muscles (ie, abductor pollicis brevis, flexor pollicis brevis, and the opponens pollicis).

There are two simple tests which can be performed in assessing the possibility of carpal tunnel syndrome—Tinel's test and Phalen's sign.

1. *Tinel's test*
   This involves lightly tapping over the carpal tunnel at the wrist joint. This tapping will reproduce the shooting sensations and exacerbate the symptoms of the numbness and parasthesia. Try not to do this too often in the examination as it can be rather painful and uncomfortable for the patient.
2. *Phalen's sign*
   Phalen's sign is where you should ask the patient to flex their wrist, or passively flex it for them, for a minute or so, or until it is uncomfortable for the patient. If it is positive it will result in the appearance of the symptoms, or worsening of them. Do not perform this assessment for too long as, again, it can be rather uncomfortable for the patient.

The causes of carpal tunnel syndrome are wide and varied and can include pregnancy, arthritis, wrist trauma, hypothyroidism, or diabetes mellitus. These conditions need to be carefully managed, but it may be necessary for immediate treatment of the carpal tunnel syndrome.

Investigation of carpal tunnel syndrome can be undertaken by a comprehensive history and examination of the wrist joint, including Tinel's test and Phalen's sign. If arthritis is considered as a cause of the carpal tunnel syndrome, an X-ray may be required to confirm this. Some practitioners like to undertake electromyograms or nerve conduction studies (Mayoclinic), but a lot can be gained from the history and examination.

Treatment of carpal tunnel syndrome can be done by providing a wrist splint for the patient, administering antiinflammatory drugs, or local steroid injections. Local steroid injections tend not to last very long, and it may be necessary to undertake surgical decompression of the carpal tunnel. This can be done by an open procedure or by endoscopic means.

3.11.1 Ulnar Nerve

The ulnar nerve originates from the seventh and eight cervical nerves, as well as the first thoracic nerve that is C7–T1. It arises from the medial cord and typically has a root which would come from the lateral root from either the median nerve, or the lateral cord. At its origin, the ulnar nerve is found sandwiched between the axillary vein and artery, anterior to teres major.

The ulnar nerve then passes medial to the axillary artery and then continues its path down the arm lying medial to the brachial artery. At approximately mid-arm level, the ulnar nerve then pierces the medial intermuscular septum and then descends with the superior ulnar collateral artery and the ulnar collateral nerve. At that site it is found on the medial head of the triceps muscle. It then passes to the posterior aspect of the medial epicondyle, often giving a small branch to the elbow joint. Then, the ulnar nerve passes into the forearm between the two heads of the flexor carpi ulnaris.

In the forearm, the ulnar nerve gives off a *dorsal branch*. This branch passes downwards and posteriorly, between the ulna and the flexor carpi ulnaris toward the medial side of the back of the hand. After giving fine branches to the back of the hand (dorsal surface), the ulnar nerve then divides into three dorsal digital nerves. These supply the little finger, ring finger, and the middle finger (sometimes to the distal interphalangeal joint) to its medial half on the dorsal aspect. The most lateral of the dorsal digital nerves communicate with the superficial branch of the radial nerve. Occasionally, a fourth digital nerve (dorsal) may be present and can extend to the index finger.

The dorsal digital branches of the radial and also the ulnar nerves do not extend all the way to the tips of the fingers. In the first and the fifth fingers, the branches do extend as far as the nail beds, but in the intermediate three fingers, the innervation typically passes to the middle phalanx, or even only to the proximal phalanx. The palmar digital branches arising from the median and the ulnar nerves then complete the rest of the digits toward the nail beds.

In the inferior portion of the forearm, the ulnar nerve can give off a *palmar branch*. This branch is a variable branch and it crosses the flexor retinaculum and innervates the skin of the medial side of the palm. The ulnar nerve then continues its journey inferiorly to enter the hand in front of the carpal tunnel. Therefore, it is outside of this tunnel, and is lateral to the pisiform and passes in front of the flexor retinaculum. The ulnar nerve is found medial to the ulnar artery, and sometimes the superficial portion of the flexor retinaculum covers these structures. Occasionally, the palmaris brevis and the pisohamate ligament may well cover the ulnar nerve and artery.

The *superficial branch of the ulnar nerve* gives a fine branch to the palmaris brevis. The superficial branch then divides into palmar digital nerves which pass to the medial side of the little finger and adjacent sides of the little and ring fingers. These branches innervate both skin and articular structures, specifically to the metacarpophalangeal and interphalangeal joints, as do those of the median nerve, and these communicate with the median nerve. Sometimes, the ulnar nerve can also supply the adjacent sides of the ring and middle fingers.

The *deep branch of the ulnar nerve* goes between the flexor digiti minimi and the abductor digiti minimi, and supplying them as it does so. The deep branch then passes through a fibrous arch in the proximal end of the opponens digit minimi, innervating it too. This branch will then wind round the hook of the hamate and then pass laterally alongside the deep palmar arch, under the cover of a fat pad behind the flexor tendons. During this journey the deep branch innervates the adductor pollicis, third and fourth lumbricals, all of the interossei and typically the flexor pollicis brevis. Due to its extensive innervation, it has been referred to as the *"musician's nerve"* as it innervates the

Table 3.8 This Table Summarizes the Branches of the Ulnar Nerve, and the Muscles, Skin and Joints That It Innervates

Ulnar nerve branches		
Dorsal branch	Twigs to back of hand	
	Dorsal digital nerves	Little finger, ring finger and middle fingers dorsally
Palmar branch	Medial side of palm	
Superficial branch	Palmaris brevis	
	Palmar digital nerves	Index finger and medial half of ring finger
		Metacarpophalangeal and interphalangeal joints
Deep branch	Flexor digiti minimi Abductor digiti minimi Adductor pollicis Third and fourth lumbricals All interossei Flexor pollicis brevis	

muscles responsible for fine movements of the fingers. In contrast, the median nerve has been referred to as the *"laborer's nerve"* as it innervates muscles typically used in those professions, not involved in fine finger movements.

In summary, the ulnar nerve innervates the palmar aspects of the medial one and a half fingers and the dorsal aspects of the medial two and a half fingers. The median nerve however, innervates the other fingers on the palmar surface, and the radial nerve innervates the dorsal aspect. The median nerve extends dorsally on the distal phalanx of the thumb and the distal two phalanges of the first two and a one-half fingers. This is summarized in Table 3.8.

The muscles innervated by the ulnar nerve are described in Table 3.9. This provides a summary of the origin, insertion, innervation, and the function of each of these muscles.

3.12 CLINICAL APPLICATIONS

When undertaking any clinical history taking or examination, you should always do the following, and follow a logical and systematic format:

a. Introduce yourself to the patient
b. Advise them of what position you hold for example student, specialty grade, consultant etc.

Table 3.9 This Table Summarizes the Muscular Innervation of the Ulnar Nerve, and Where Each Muscle Originates from, Where It Inserts, and Also Describes the Function of Each Muscle

Muscle	Origin	Insertion	Innervation	Function
Palmaris brevis	Palmar aponeurosis Flexor retinaculum	Skin on ulnar side of the palm	Superficial branch of the ulnar nerve	Tension of the skin of the palm Deepens the hollow of the palm
Flexor digiti minimi (brevis)	Hook of hamate	Medial side of base of proximal phalanx of the fifth digit	Deep branch of the ulnar nerve	Flexion of the fifth (little) digit
Abductor digiti minimi	Pisiform	Medial side of base of proximal phalanx of the fifth digit	Deep branch of the ulnar nerve	Abduction of fifth (little) digit
Adductor pollicis	Oblique head – Anterior aspect of the base of the second metacarpal, capitate and the trapezoid Transverse head – third metacarpal	Medial sesamoid bone Medial side of the base of the proximal phalanx of thumb Extensor expansion	Deep branch of the ulnar nerve	Adduction of the thumb Aids opposition
Lumbricals (3 and 4)	Flexor digitorum profundus	Extensor expansion	Deep branch of the ulnar nerve (N.B. The first and second lumbricals are innervated by the median nerve)	Flexion of the metacarpo-phalangeal joints Extension of the inter-phalangeal joints
Flexor pollicis brevis	Trapezium Flexor retinaculum	Proximal phalanx of the thumb	Deep branch of the ulnar nerve for its medial side; recurrent branch of the median nerve	Flexion of the thumb at the first metacarpo-phalangeal joint

c. Your reason for consulting with them, or to find out why they have presented to you

d. Always take a thorough and detailed history, which will be guided by the presenting signs and symptoms

e. When examining the patient, always tell them what you will ask them to do, or what region of the body you will be examining, with specific instructions and ensure they give consent

A detailed examination and history taking should be completed as described in Chapter 1.

The following table (Table 3.10) summarizes the nerve roots, muscle supplied, and relevant clinical tests to ask the patient to perform in assessing the ulnar nerve.

Table 3.10 This Demonstrates the Origin of Each of the Vertebral Levels of the Ulnar Nerve, the Muscles They Supply. It Also States What to Ask the Patient to Do When Assessing Each Muscle and Therefore Approximate Nerve Root of the Ulnar Nerve

Nerve root	Muscle	Ask patient to...
C7-8	Flexor carpi ulnaris	Abduct little finger and identify the tendon when all the fingers are fully extended
C8-T1	Flexor digitorum profundus (digits 3 and 4)	Fix the middle phalanx of the little finger and resist extension to the distal phalanx
C8-T1	Dorsal interossei	Abduction of the fingers
C8-T1	Palmar interossei	Adduction of the fingers
C8-T1	Adductor pollicis	Adduction of the thumb
C8-T1	Abductor digiti minimi	Abduction of the little finger
C8-T1	Opponens digit minimi	With all the fingers extended, bring the little finger over the other fingers

As well as the assessment of the muscles that the ulnar nerve supplies, sensation should also be assessed in the distribution of the ulnar nerve that is the index finger and medial half of the ring finger. Again, this should be according to the International Standards for Neurological Classification of Spinal Cord Injury (ISNCSCI) as previously described in Chapter 1.

3.13 CLINICAL APPLICATIONS

1. *Ulnar nerve injury*
   Injury to the ulnar nerve tends to happen in one of the four key areas:
   a. Behind the medial epicondyle of the humerus
   b. Cubital tunnel that is a tunnel bordered by the medial epicondyle of the humerus, olecranon of the ulna, and the tendinous arch formed by the ulnar and humeral heads of the flexor carpi radialis
   c. Wrist joint
   d. Hand
   Typically, injury to the ulnar nerve happens as it runs behind the medial epicondyle of the humerus, and can occur with a fracture at that site.
   Injury to the ulnar nerve will present with numbness and parasthesia in the little finger and the medial half of the ring finger and also the medial side of the palm. As well as the sensory deficit,

there can be extensive motor loss to the hand. This will result in reduced wrist adduction when trying to flex the wrist, and the hand will be drawn over to the radial side due to the unopposed pull of the flexor carpi radialis, which is innervated by the median nerve. This is due to a lack of innervation of the flexor carpi ulnaris by the ulnar nerve. The patient will also have difficulty making a fist because there is a lack of opposition, and the meta-carpophalangeal joints will become hyperextended and unable to flex the fourth and fifth digits at the distal interphalangeal joints. In addition to this, the patient will be unable to extend the inter-phalangeal joints when trying to straighten the digits. The charac-teristic appearance of the patient's hand will result in a "claw hand" due to wasting of the interossei. Therefore, the "claw hand" appearance is because of the unopposed action of the extensors and also the flexor digitorum profundus.

2. *Cubital tunnel syndrome*

   Tunnel is created by the ulnar and humeral heads of the flexor carpi ulnaris. If the ulnar nerve is compressed at that point, it is referred to as cubital tunnel syndrome. This syndrome is caused because of sleeping with the arms behind the neck, and the elbow joint flexed, or pressing the elbows when in the typing position for office work-ers, or indeed from strenuous bench-press exercises. The patient will complain of pain and numbness (parasthesia) along the sensory dis-tribution of the ulnar nerve that is the medial one and a half fingers as well as the medial portion of the palm. It may be treated by con-servative measures like removing or alleviating the exacerbating fac-tor, and may require surgery.

3. *Guyon's tunnel syndrome*

   Guyon's canal or tunnel, or the ulnar canal or tunnel is a canal formed at the wrist joint by the following structures that allow the ulnar nerve and artery a route to pass into the hand:
   a. Superior boundary—Superficial palmar carpal ligament
   b. Floor—Hypothenar muscles and the deep flexor retinaculum
   c. Medial boundary—Pisiform and pisohamate ligament
   d. Lateral boundary—Hook of the hamate

Guyon's tunnel syndrome results in the patient complaining of reduced sensation or numbness and parasthesia of the medial one and a half fingers and weakness of the intrinsic muscles of the hand. Clawing of the hand may be present with hyperextension of the metacarpophalangeal joints with flexion at the proximal interphalangeal joint affecting the fourth and fifth digits. Flexion is not affected and radial deviation is not present.

Shea and McClain (1969) divided pathologies affecting the ulnar nerve into three types. Type I involved pathology affecting the ulnar nerve proximal to or within Guyon's canal. This would exhibit both motor and sensory abnormalities. Type II affected only the deep branch of the ulnar nerve and resulted in weakness in muscles innervated by that branch, and could spare the hypothenar muscles. Type III involves compression of the ulnar nerve at the end of Guyon's canal, resulting in only a sensory deficit and no motor-related problems.

Typical problems to affect the ulnar nerve and result in Guyon's tunnel syndrome arise from a ganglion, trauma, musculotendinous arch, or disease of the ulnar artery, and surgery may be offered to these patients for alleviating pain, numbness, and muscle weakness (Aguiar et al., 2001).

3.13.1 Radial Nerve

The radial nerve arises from the posterior cord of the brachial plexus. As the largest branch of the brachial plexus, it courses behind the third part of the axillary artery and also the superior part of the brachial artery. As it passes behind the brachial artery it soon dips forward with the profunda brachii artery. It passes anterior to the tendons of teres major and latissimus dorsi and runs in front of the subscapularis. With the profunda brachii artery, and later on, its radial collateral branch, the radial nerve then passes around the humerus obliquely. It then passes between the lateral and medial heads of the triceps brachii, and then within a shallow groove deep to the lateral head of the triceps.

Once the radial nerve reaches the outer aspect of the humerus, it pierces the intermuscular septum (lateral) and then passes to the anterior compartment of the arm. It then passes sandwiched between the brachialis medially and the extensor carpi radialis inferolaterally and the brachioradialis superolaterally. At or below the level of the lateral epicondyle, the radial nerve then divides into its superficial and deep branches.

In terms of surface anatomy, the radial nerve extends from the medial margin of the biceps, opposite the posterior axillary fold, obliquely across the back of the arm. It pierces the intermuscular septum at the superior point of trisection of a line between the deltoid insertion and the lateral epicondyle and then descends to the front of the lateral epicondyle. The radial nerve can be palpated in thin individuals as it winds around the humerus. Typically, it can be felt in these individuals about 1−2 cm inferior to the deltoid insertion, and also in the interval between the brachialis and the brachioradialis.

In the arm the radial nerve can be found if a line is imaginarily drawn from the start of the brachial artery across the elevations from the lateral and long heads of the triceps brachii muscle to the junction of the upper and middle thirds if a line is drawn from the deltoid tuberosity to the lateral epicondyle. Again, the radial nerve can also be found approximately 10 mm lateral to the tendon of the biceps brachii muscle.

The radial nerve has the following branches: *cutaneous, articular, muscular, and the superficial and deep terminal branches.*

1. *Cutaneous branches*
   The cutaneous nerves of the radial nerve are the lower lateral cutaneous nerve of the arm and the posterior cutaneous nerve of the arm.
   The lower lateral (brachial) cutaneous nerve of the arm pierces the lateral head of the triceps brachii muscle inferior to the insertion of the deltoid muscle. The lower lateral cutaneous nerve of the arm then courses anterior to the elbow, closely related to the cephalic vein and innervates the skin of the lower portion of the inferior half of the arm.
   The posterior (brachial) cutaneous nerve of the arm is smaller and originates in the axilla and courses to the inner aspect of the arm innervating the skin of the dorsal aspect of the arm almost to the olecranon. It also passes behind and communicates with, the inter-costobrachial nerve. The intercostobrachial nerve is a cutaneous branch of the intercostal nerve.
2. *Articular branches*
   The articular branches of the radial nerve innervate the elbow joint.

3. *Muscular branches*

The muscular branches of the radial nerve innervate the extensor carpi radialis longus, brachialis, brachioradialis, anconeus, and the triceps. These are grouped as lateral, medial, and posterior.

The *lateral muscular branches* originate from the nerve as it is in front of the lateral intermuscular septum. The lateral muscular branches innervate the extensor carpi radialis longus, brachioradialis, and the brachialis.

The *medial muscular branches* originate from the radial nerve on the inner aspect of the arm and innervate the long and medial heads of the triceps. The branch that innervates the medial head of the triceps is a long thin branch which is positioned very close to the ulnar nerve down to the distal one-third of the arm and can also be termed the *ulnar collateral nerve*.

The posterior muscular branch is larger and arises from the radial nerve as it is positioned within the groove. It innervates the anconeus, as well as the lateral and medial heads of the triceps muscle. The branch to the anconeus is a long nerve and this branch is within the substance of the medial head of the triceps muscle, and also innervates it too. This branch is also accompanied by the profunda brachii artery, and it goes posterior to the elbow joint and then terminates in the anconeus.

4. *Superficial branch*

The superficial or superficial terminal branch passes inferior and anterior to the lateral epicondyle and along the outer aspect of the upper two-thirds of the forearm. First it lies on supinator, lateral to the radial artery and posterior to the brachioradialis. It then is found posterior to the brachioradialis in the middle one-third of the forearm and runs close to the lateral side of the radial artery. It is then found on the pronator teres then on the flexor digitorum superficialis (radial side) and then positioned on the flexor pollicis longus. It then moves away from the radial artery approximately 7 cm above the wrist. The superficial terminal branch then passes under the tendon of the brachioradialis and swings around the lateral aspect of the radius as it passes distally and it goes through the deep fascia and terminates in four or five digital nerves.

5. *Dorsal digital nerves*

The dorsal digital nerves are small and there can be either four or five of these. The dorsal digital nerves frequently join with the lateral and posterior cutaneous nerves (mentioned earlier). Note that the digital nerves innervating the thumb only pass as far as the root of the nail. The dorsal digital nerves innervating the middle finger only reach the middle phalanx and those to the middle and ring finger do not pass any further than the proximal interphalangeal joints. The rest of the distal aspect of the skin on the dorsal aspect of the digits are innervated by the median and ulnar nerves palmar digital nerves.

The following table summarizes what the branches supply (Table 3.11).

Table 3.11 This Summarizes the Innervation Territories of the Dorsal Digital Nerves	
Dorsal digital nerve	Innervation
1st	Skin of radial aspect of the thumb and adjacent part of the thenar eminence
2nd	Medial aspect of the thumb
3rd	Lateral aspect of the index finger
4th	Adjoining side of index and middle finger
5th	Adjoining aspect of the middle and ring finger

6. *Deep terminal branch of the radial nerve*

The deep terminal branch of the radial nerve is also referred to as the posterior interosseous nerve. The deep branch of the radial nerve is muscular and articular in its distribution. It originates from under the brachioradialis and passes laterally around the radius, sandwiched between the superficial and deep layers of the supinator, which it also innervates. It typically will contact the bare surface of the radius and can be susceptible to damage if there is a fracture of the radius at this site. When the deep terminal branch of the radial nerve reaches the back of the forearm, it is found between the superficial and deep extensors, giving innervation to the superficial group, and is accompanied by the posterior interosseous artery. For the rest of the course of tuso nerve, it is termed the posterior interosseous nerve, a name that has also been applied to the deep terminal branch in its entirety. In the distal part of the forearm, it passes onto the interosseous membrane by going deep to the extensor pollicis longus. It then is found with the anterior interosseous artery, in the groove for the extensor digitorum on the back of the radius. It terminates on the back of the carpus in an enlargement from which twigs are sent to the wrist joint and the intercarpal joints.

As the deep terminal branches of the radial nerve pass through the forearm, they supply the supinator (and frequently the extensor carpi radialis brevis), the extensor digitorum, extensor digiti minimi, and the extensor carpi ulnaris. The posterior interosseous nerve innervates the abductor pollicis longus, extensor pollicis brevis, extensor pollicis longus as well as the extensor indicis.

A summary of the branches of the radial nerve branches is given in Table 3.12. The muscles innervated by the radial nerve are summarized in terms of their origin, insertion, actions, and nerve supply in Table 3.13.

Table 3.12 This Provides a Summary of the Branches of the Radial Nerve and the Motor, Sensory, or Articular Innervations of Each Branch

Branch	Innervation
Cutaneous	
Lower lateral	Skin of lower portion of the inferior half of the arm
Posterior	Dorsal aspect of arm (to olecranon)
Articular	Elbow joint
Muscular	
Lateral	Extensor carpi radialis longus Brachioradialis Brachialis
Medial	Long and medial heads of the triceps
Posterior	Lateral and medial heads of the triceps muscle Anconeus
Superficial	Terminates in the dorsal digital nerves (see below)
Dorsal digital nerves	
1st	Skin of radial aspect of the thumb and adjacent part of the thenar eminence
2nd	Medial aspect of the thumb
3rd	Lateral aspect of the index finger
4th	Adjoining side of index and middle finger
5th	Adjoining aspect of the middle and ring finger
Deep terminal branch	Supinator Extensor carpi radialis brevis Extensor digitorum Extensor digiti minimi Extensor carpi ulnaris Abductor pollicis longus Extensor pollicis brevis Extensor pollicis longus Extensor indicis

Table 3.13 This Highlights Each Muscle That Is Supplied by Specific Branches of the Radial Nerve (Including Vertebral Level Origin), Its Origin, Insertion, and Functions

Muscle	Origin	Insertion	Actions	Innervation
Brachialis	Distal two-thirds of the anteromedial and anterolateral surface of the humerus	Capsule of the elbow joint Coronoid process Tuberosity of the ulna	Flexion of the forearm	Musculocutaneous nerve (C5-6) and a small contribution from the radial nerve (C7) to its lateral aspect
Brachioradialis	Upper part of the lateral supracondylar ridge of the humerus	Lateral surface of the radius (just superior to the styloid process)	Flexion of the forearm when mid-pronated	Radial nerve (C5-6)
Triceps	Long head – Infraglenoid tubercle of the scapula Lateral head – Posterior surface of the humerus above grove for radial nerve Medial head – Posterior surface of the humerus below grove for radial nerve	Posterior portion of the upper aspect of the olecranon Fascia of the forearm via the tricipital aponeurosis	Extension of the forearm Lateral and long heads recruited when resistance occurs Long head can also hold the humerus in place during its actions	Radial nerve (C6-8)
Anconeus	Lateral epicondyle of humerus	Olecranon and Posterior surface of ulna	Assists triceps in extension of the forearm Active during pronation	Radial nerve (C6-8)
Extensor carpi radialis longus	Lower portion of the lateral supracondylar ridge	Back of the base of the 2nd metacarpal	Extension of the hand Abduction of the hand at the wrist joint	Radial nerve (C6-7)
Supinator	Superficial part - Lateral epicondyle of the humerus Deep part – Supinator fossa and crest Oblique line of ulna	Superficial part – Radius Deep part – Upper third of shaft of the radius	Supination of the forearm	Posterior interosseous nerve (C6-7)

Muscle	Origin	Insertion	Action	Innervation
Extensor carpi radialis brevis	Lateral epicondyle of humerus	Back of the bases of the 2nd and 3rd metacarpals	Extension of the hand Abduction of the hand at the wrist joint	Deep branch of radial nerve (C7-8)
Extensor digitorum	Lateral epicondyle of humerus	Extensor expansion and the middle and distal phalanges of fingers 2-5	Extends the proximal phalanges on the metacarpals	Posterior interosseous nerve (C7-8)
Extensor digiti minimi	Lateral epicondyle of humerus	Extensor aponeurosis of the little finger	Extension of the proximal phalanx of the little finger	Posterior interosseous nerve (C7-8)
Extensor carpi ulnaris	Lateral epicondyle of humerus	Tubercle on medial side of base of 5th metacarpal	Extends hand and acts with the radial extensors. Pure adduction when acting with flexor carpi ulnaris	Posterior interosseous nerve (C7-8)
Abductor pollicis longus	Interosseous membrane Posterior surface of the radius and ulna	Lateral aspect of the base of the first metacarpal and (typically) the trapezium	Abduction of the first metacarpal at the carpometacarpal joint Stabilizes the first metacarpal during movement of the phalanges	Posterior interosseous nerve (C7-8)
Extensor pollicis brevis	Posterior surface of the radius	Back of the proximal phalanx of thumb	Extension of the thumb at metacarpophalangeal joint	Posterior interosseous nerve (C7-8)
Extensor pollicis longus	Posterior surface of ulna (middle) Interosseous membrane	Base of distal phalanx of thumb (dorsal aspect)	Extension of distal phalanx via extension of the interphalangeal joint of the thumb	Posterior interosseous nerve (C7-8)
Extensor indicis	Posterior surface of the ulna Interosseous membrane	Extensor expansion of the index finger	Extension of the proximal phalanx of the index finger	Posterior interosseous nerve (C7-8)

3.14 CLINICAL APPLICATIONS

When undertaking any clinical history taking or examination, you should always do the following, and follow a logical and systematic format:

a. Introduce yourself to the patient
b. Advise them of what position you hold for example student, specialty grade, consultant etc.
c. Your reason for consulting with them, or to find out why they have presented to you
d. Always take a thorough and detailed history, which will be guided by the presenting signs and symptoms
e. When examining the patient, always tell them what you will ask them to do, or what region of the body you will be examining, with specific instructions and ensure they give consent

A detailed examination and history taking should be completed as described in Chapter 1.

The following summary table (Table 3.14) will highlight what nerve roots of the radial nerve have to be tested, what muscle it supplies, a brief description of the normal function, and how to test for it clinically.

3.15 CLINICAL APPLICATIONS

Damage to the radial nerve may happen at several points along its course. Typically, the radial nerve may be damaged by trauma or entrapment of the nerve.

1. *Fracture of the shaft of the humerus*
   If the humerus is fractured at mid-humerus level, the radial nerve could be injured in the radial grove (or radial sulcus) of the humerus. If there is a fracture at this site, it tends not to affect the triceps as the nerve supply to two of the three heads of the triceps muscle tends to arise more proximal to this point. However, there may be weakness of the triceps muscle. However, muscles in the posterior compartment of the forearm may become weakened or paralyzed. This would result in paralysis of the brachioradialis, supinator, and also the extensor muscles within the fingers and hand. This would result in the clinical presentation of *wrist-drop*.

Table 3.14 This Summarizes the Nerve Roots That Contribute to the Radial Nerve, What Specific Musculature They Supply, the Typical Functions of Those Muscles, and the Clinical Test to Assess Function of That Nerve Root and Muscle

Nerve root	Muscle	Function	Testing
C5-6	Brachioradialis	Flexion of the foreram when mid-pronated	Have the patient flex elbow midway between pronation and supination
C6-7	Supinator	Supination	
C6-7	Extensor carpi radialis longus	Extends and abducts the wrist	Extension of the pateint's wrist to the side of the radius with fingers in the extended position
C6-8	Triceps brachii	Extension of the forearm at elbow joint	Exted the patients elbow against resistance
C7-8	Extensor digitorum	Extension of the index, middle, ring and little fingers; extension of the wrist	Ask patient to maintain their fingers in the extended position at the metacarpophalangeal joints
C7-8	Extensor carpi ulnaris	Extension and abduction of the wrist joint	Ask patient to extend their writs to the ulnar side
C7-8	Abductor pollicis longus	Abduction of the first metacarpal at the carpometacarpal joint	Ask patient to abduct their thumb
C7-8	Extensor pollicis brevis	Extension of the thumb at metacarpophalangeal joint	Ask patient to extend their thumb at the metacarpophalangeal joint
C7-8	Extensor pollicis longus	Extension of distal phalanx via extension of the interphalangeal joint of thumb	Ask the patient to resist flexion of the thumb at the interphalangeal joint of the first digit

This is seen clinically with the inability to extend the fingers and wrist joint at the metacarpophalangeal joints. The "relaxed" wrist joint becomes more flexed due to the unopposed contraction of the flexor muscles of the wrist and fingers, hence the *wrist-drop*. Palsy of the radial nerve has been said to be the most common nerve lesion affecting the long bones due to fracture (Rockwood et al., 1996).

2. *Radial nerve injury at the elbow*

   There may be compression of the radial nerve at the elbow due to thickened fibrous tissue proximal to the radial tunnel. There may also be pronounced tissue at the upper edge of the superficial part of the supinator resulting in compression of the posterior interosseous nerve—a continuation of the deep branch of the radial nerve. This is referred to as the *arcade of Frohse*, named after the German anatomist Fritz Frohse who described it.

   Radial nerve compression may also be caused by the recurrent radial artery, referred to as hypertrophic leash of Henry. This results in compression of the deep branch of the radial nerve and can present with chronic forearm pain (Loizides et al., 2011).

3. *Radial nerve injury at the wrist and hand*

   The distal branches of the radial nerve may be affected by fractures of the distal radius, soft tissue mass at the wrist, or by a prominent extensor carpi radialis brevis compressing the nerve. Lesions to the superficial branches of the radial nerve could arise from a tight plaster cast at that site, handcuffs, or tight wristbands. This would result in pain and perhaps anesthesia along the distribution of the sensory branching of the radial nerve but would not result in it affecting the muscles of the hand.

3.15.1 Axillary Nerve

The axillary nerve arises from the fifth and the sixth cervical nerves that is C5–6. The axillary nerve is a branch of the *posterior cord*. It is found anterior to the subscapularis, posterior to the brachial artery, and lateral to the radial nerve. At the inferior end of the subscapularis the axillary nerve then runs posterior, close to the joint capsule, passing through the quadrangular space with the posterior circumflex humeral artery, sandwiched between the lateral and long heads of the triceps muscle. The axillary nerve is found inferior to the capsule of the shoulder joint, and it sends a small branch to this joint. The

Table 3.15 This Table Summarizes the Muscles Which Are Innervated by the Axillary Nerve in Terms of Their Origin, Insertion, and Actions

Muscle	Origin	Insertion	Innervation	Function
Deltoid	Front of superior surface of the lateral third of the clavicle, adjoining acromion, spine of the scapula	Deltoid tuberosity of the humerus	Axillary nerve	Acromial part – powerful abductor of arm Spinous part – extension of arm and lateral rotation Clavicular part – flexion of arm and medial rotation
Teres minor	Lateral margin of the infraspinous fossa	Capsule of the shoulder joint Lowest facet on the greater tubercle of the humerus	Axillary nerve	Lateral rotation of the arm Maintains the head of the humerus in place during abduction

axillary nerve then winds medial to the surgical neck of the humerus, and is typically in contact with this part of the bone. The axillary nerve then usually divides into two branches at this point under the cover of the deltoid muscle. The *anterior branch of the axillary nerve* winds round the humerus deep to the deltoid muscle, and also innervates the muscle at this point. The *posterior branch of the axillary nerve* innervates the teres minor and also the deltoid. The posterior branch then winds round the deltoid muscle and goes to innervate an area of skin on the back of the arm as the upper lateral brachial cutaneous nerve, or the superior lateral cutaneous nerve of arm. This nerve innervates specifically the skin over the inferior two-thirds of the deltoid muscle at its posterior aspect. Superior to this, the skin of the shoulder is innervated by the supraclavicular nerves. The level of the axillary nerve can also be thought of as a horizontal plane through the middle of the deltoid muscle. Table 3.15 explains more about the muscles innervated by the axillary nerve.

3.16 CLINICAL EXAMINATION

When undertaking any clinical history taking or examination, you should always do the following, and follow a logical and systematic format:

a. Introduce yourself to the patient
b. Advise them of what position you hold for example student, specialty grade, consultant etc.

c. Your reason for consulting with them, or to find out why they have presented to you
d. Always take a thorough and detailed history, which will be guided by the presenting signs and symptoms
e. When examining the patient, always tell them what you will ask them to do, or what region of the body you will be examining, with specific instructions and ensure they give consent

A detailed examination and history taking should be completed as described in Chapter 1.

Specific to the axillary nerve, the following should be examined:

1. *Assessment of the deltoid*
   This can be undertaken by asking the patient to fully abduct their arm. Symmetry over the deltoid muscle should be noted and the left and right sides compared.
2. *Assessment of the teres minor*
   The function and power of the teres minor can be assessed by the *Hornblower's test*. The upper limb should be abducted by the examiner out to 90° in the plane of the scapula, then flex the elbow. The examiner should place medial pressure at the elbow and hold the hand in position whilst asking the patient to perform external (lateral) rotation against resistance by the examiner. A positive Hornblower's test would result in pain and/or dysfunction during this procedure. The other way to do this is by asking the patient to bring their hands toward their mouth, but without abducting the shoulder joint. Therefore, this test assesses the posterior part of the rotator cuff muscles. The teres minor is required to bring the patient's hands toward their mouth and stabilize the lateral rotation of the shoulder joint.
3. Sensory assessment
   The axillary nerve also provides innervation to skin over the inferior two-thirds of the deltoid muscle at its posterior aspect. This is referred to as the regimental badge area. Injury to the axillary nerve will result in reduced or absent sensation over this anatomical territory.

Nerve	Components	Functions	Point of Entry/ Exit from Brain	Exits / enters cranial cavity	Nuclei	Ganglion	Important Branches
Olfactory (I)	Special sensory	Smell	Forebrain	Cribriform plate of ethmoid bone	No specific nucleus. Olfactory epithelium contain the cell bodies	None	Olfactory epithelium (central processes)
Optic (II)	Special sensory	Vision	Midbrain	Optic canal	Lateral geniculate nucleus	Retinal ganglion cells	Optic nerve; optic tract
Oculomotor (III)	Somatic motor	Extra-ocular muscles	Midbrain	Superior orbital fissure	Oculomotor nucleus; Edinger-Westphal nucleus	Ciliary ganglion	Motor branches to extra-ocular muscles;
	Visceral motor	Sphincter muscle and ciliary muscle					Parasympathetic division
Trochlear (IV)	Somatic motor	Innervates the superior oblique muscle	Midbrain	Superior orbital fissure	Nucleus of the trochlear nerve	None	None. Only supplies the superior oblique muscle
Trigeminal (V)	General sensory; Branchial motor	Sensation from face; paranasal sinuses; nose and teeth Muscles of mastication	Pons	Superior orbital fissure (Va), foramen rotundum (Vb) or foramen ovale (Vc)	Spinal trigeminal nucleus; Pontine trigeminal nucleus; Mesencephalic trigeminal nucleus; Trigeminal motor nucleus	Trigeminal ganglion; Submandibular ganglion	Ophthalmic nerve; Maxillary nerve; Mandibular nerve
Abducent (VI)	Somatic motor	Innervates the lateral rectus muscle	Pontomedullary junction	Superior orbital fissure	Abducent nerve nucleus	None	None. Only supplies the lateral rectus muscle

(Continued)

Nerve	Components	Functions	Point of Entry/ Exit from Brain	Exits / enters cranial cavity	Nuclei	Ganglion	Important Branches
Facial (VII)	Branchial motor	Muscles of facial expression, stylohyoid, stapedius, posterior belly of digastric	Pontomedullary junction	Stylomastoid foramen	Facial motor nucleus; lacrimal nucleus; superior salivatory nucleus; Gustatory nucleus; Spinal trigeminal nucleus	Geniculate ganglion; Pterygopalatine ganglion; Submandibular ganglion	**Intra-temporal** Greater petrosal nerve; nerve to stapedius; chorda tympani **Extra-temporal** Temporal; zygomatic; buccal; marginal mandibular; cervical; posterior auricular; posterior belly of digastric branch; stylohyoid branch
	Visceral motor	Parasympathetic innervation of the submandibular and sublingual salivary glands, lacrimal gland and the nasal and palatal glands					
	Special sensory	Anterior two-thirds of the tongue (and palate)					
	General sensory	Concha of the auricle					
Vestibulo-cochlear (VIII)	Special sensory	Balance for the vestibular component; Hearing for the spiral (cochlear) component	Pontomedullary junction	Internal auditory meatus	Vestibular nucleus; ventral cochlear nucleus; dorsal cochlear nucleus; superior olivatory nucleus	Vestibular ganglion; Spiral ganglion	Vestibular nerve; cochlear nerve
Glosso-pharyngeal (IX)	Branchial motor	Stylopharyngeus	Medulla oblongata	Jugular foramen	Nucleus ambiguus; Solitary nucleus; Spinal trigeminal nucleus; Inferior salivatory nucleus	Inferior ganglion; Otic ganglion; Superior ganglion; Inferior ganglion	Muscular; Tympanic; Pharyngeal; Tonsillar; Carotid sinus branch
	Visceral motor	Parotid gland for parasympathetic innervation					
	Special sensory	Taste from the posterior one-third of the tongue					
	General sensory	External ear					
	Visceral sensory	Pharynx; parotid gland; middle ear; carotid sinus and body					

Nerve	Component	Location	Exit	Nucleus	Ganglion	Branches
Vagus (X)	Branchial motor	Pharyngeal constrictors; Laryngeal muscles (intrinsic); Palatal muscles; Upper two-thirds of oesophagus	Medulla oblongata	Jugular foramen	Dorsal nucleus of the vagus nerve; Nucleus ambiguus; Soliatry nucleus; Spinal trigeminal nucleus	Superior ganglion; Inferior ganglion	Meningeal branch; Auricular branch; Pharyngeal branches; Superior laryngeal nerve; Recurrent laryngeal nerve; carotid branches; Cardiac branches; Oesophageal branches; Pulmonary branches; Gastric branches; Coeliac branches; Renal branches
	Visceral motor	Heart; Trachea and bronchi; Gastrointestianl tract					
	Special sensory	Taste from the palate and the epiglottis					
	General sensory	Auricle; External auditory meatus; Posterior cranial fossa dura mater					
	Visceral sensory	Gastrointestinal tract (to last one-third of the transverse colon); Pharynx and Larynx; Trachea and Bronchi; Heart					
Spinal Accessory (XI)	Somatic motor	Innervates the sternocleido-mastoid and trapezius muscles	Medulla oblongata (and spinal cord)	Jugular foramen	Nucleus ambiguus; Spinal accessory nucleus	None	Cranial branch; Spinal branch
Hypoglossal (XII)	Somatic motor	Extrinsic and intrinsic muscles of the tongue. Palatoglossus is not supplied by the hypoglossal nerve. It is supplied by the glossopharyngeal nerve	Medulla oblongata	Hypoglossal canal	Hypoglossal nucleus	None. It may however receive general sensory fibres from the ganglion of C2	Meningeal branches; Thyrohyoid branches; Muscular branches;

3.17 CLINICAL APPLICATIONS

1. *Fractures of the surgical neck of the humerus*

   The axillary nerve can be damaged due to a fracture of the surgical neck of the humerus as the nerve winds around this point, just inferior to the head of the humerus. As well as fractures to the surgical neck of the humerus, the axillary nerve may also be damaged by shoulder (glenohumeral) dislocation or from crutches pressing in the axilla through improper use or provision.

   On examination, the patient would have a flattened deltoid and also there may be pitting over the acromion. There will be sensory loss as previously described.

2. *Intramuscular injection to deltoid*

   The deltoid may be used as a muscle to give intramuscular injections too. The clinician should be aware of the axillary nerve running under the deltoid at the surgical neck of the humerus. This should therefore prevent injury to this nerve.

3. *Axillary brachial plexus block*

   The axillary brachial plexus block is typically performed for hand and forearm surgery, and should be undertaken using ultrasound guidance. This technique of anaesthetizing the brachial plexus is considered superior compared to supraclavicular or interscalene blocks. The anesthesia extends from the mid-arm level down to the hand. Although it can be referred to as the axillary brachial plexus block, this is only due to the access to the brachial plexus via the axilla and does NOT anesthetize the axillary nerve due to its origin from the posterior cord high up within the axilla (NYSORA).

REFERENCES

Aguiar, P.R., Bor-Seng-Shu, E., Gomes-Pinto, F., Almeida-Leme, R.J., Freitas, A.B.R., Martins, R.S., et al., 2001. Surgical management of Guyon's canal syndrome. Arq. Neuropsiuiatar. 59 (1), 106–111.

Allen, Z.A., Shanahan, E.M., Crotty, M., 2010. Does suprascapular nerve block reduce shoulder pain following stroke: a double-blind randomized controlled trial with masked outcome assessment. BMC Neurol. 10, 83–87.

Canadian Orthopaedic Trauma Society, 2007. Nonoperative treatment compared with plate fixation of displaced midshaft clavicular fractures. A multicenter, randomized clinical trial. J. Bone Joint Surg. Am. 89 (1), 1–10.

Fernandes, M.R., Fernandes, R.J., 2010. Artroscopia no tratamento da tendinite calcária refratária do ombro. Rev. Bras. Ortop. 45, 53–60.

Gelberman, R.H., Verdeck, W.N., Brodhead, W.T., 1975. Supraclavicular nerve-entrapment syndrome. J. Bone Joint Surg. Am. 57 (1), 119.

Gloster Jr., H.M., 2008. Complicatiosn in Cutaneous Surgery. Springer, NY, USA, p. 30.

Hempel, V., van Finck, M., Baumgartner, E., 1981. A longitudinal supraclavicular approach to the brachial plexus for the insertion of plastic cannulas. Anesth Analg. 60, 352−355.

International Standards for Neurological Classification of Spinal Cord Injury (ISNCSCI). American Spinal Injury Association. <http://www.asia-spinalinjury.org/elearning/ASIA_ISCOS_high.pdf> (accessed 05.10.15.).

Jeray, K.J., 2007. Acute midshaft clavicular fracture. J. Am. Acad. Orthop. Surg. 15 (4), 239−248.

Kothari, D., 2003. Suraclavicular brachial plexus block: a new approach. Indian J. Anaesth. 47, 2878.

Landers, J.T., Maino, K., 2012. Clarifying Erb's point as an anatomic landmark in the posterior cervical triangle. Dermatol. Surg. 38 (6), 954−957.

Loizides, A., Peer, S., Ostermann, S., Henninger, B., Stampfer-Kountchev, M., Gruber, H., 2011. Unusual functional compression of the deep branch of the radial nerve by a vascular branch (leash of Henry): ultrasonographic appearance. Rofo. 183 (2), 163−166.

Mayoclinic. <http://www.mayoclinic.org/diseases-conditions/carpal-tunnel-syndrome/basics/tests-diagnosis/con-20030332> (accessed 05.10.15.).

Nathe, T., Tseng, S., Yoo, B., 2011. The anatomy of the supraclavicular nerve during surgical approach to the clavilcuar shaft. Clin. Orthop. Relat. Res. 469, 890−894.

Nguyen, H.C., Fath, E., Wirtz, S., Bey, T., 2007. Transscalene brachial plexus block: a new posterolateral approach for brachial plexus block. Anesth. Analg. 105, 872−875.

NYSORA. The New York School of Regional Anesthesia. <http://www.nysora.com/techniques/ultrasound-guided-techniques/upper-extremity/3017-ultrasound-guided-axillary-brachial-plexus-block.html> (accessed 05.10.15.).

Pearson, A., 1937. The spinal accessory nerve in human embryos. J. Comp. Neurol. 68, 243−266.

Pearson, A.A., Sauter, R.W., Herrin, G.R., 1964. The accessory nerve and its relation to the upper spinal nerves. Am. J. Anat. 114, 371−391.

Rea, P., 2015a. Essential Clinically Applied Anatomy of the Peripheral Nervous System in the Limbs. Elsevier Academic Press, ISBN 9780128030622.

Rea, P., 2015b. Essential Clinical Anatomy of the Nervous System, first ed. Elsevier Academic Press, ISBN 9780128020302.

Rockwood Jr, C.A., Green, D.P., Bucholz, R.W., Heckman, J.D., 1996. Rockwood and Green's Fracture in Adults, fourth ed. Lippincott-Raven Publishers, Philadelphia, pp. 1043−1045.

Shea, J.D., McClain, E.J., 1969. Ulnar-nerve compression syndromes at and below the wrist. J. Bone Joint Surg. 51, 1095−1103.

Shen, W.J., Liu, T.J., Shen, Y.S., 1999. Plate fixation of fresh displaced midshaft clavicle fractures. Injury 30 (7), 497−500.

Tubbs, R.S., Benninger, B., Loukas, M., Gadol, A.A.C., 2014. Cranial roots of the accessory nerve exist in the majority of adult humans. Clin. Anat. 27, 102−107.

Note: Page numbers followed by "*f*" and "*t*" refer to figures and tables, respectively.

Printed in the United States
By Bookmasters